AAPS Introductions in the Pharmaceutical Sciences

Series Editor

Robin M. Zavod
Midwestern University
Downers Grove, IL, USA

Springer and the American Association of Pharmaceutical Scientists (AAPS) have partnered again to produce a second series that juxtaposes the *AAPS Advances in the Pharmaceutical Sciences* series. It is a set of introductory volumes that lay out the foundations of the different established pockets and emerging subfields of the pharmaceutical sciences. Springer and the AAPS aim to publish scholarly science focused on general topics in the pharma and biotech industries, and should be of interest to students, scientists, and industry professionals.

More information about this series at http://www.springer.com/series/15769

Anthony J. Hickey • Hugh D.C. Smyth

Pharmaco-complexity

Non-Linear Phenomena and Drug Product Development

Second Edition

Anthony J. Hickey
RTI International
Research Triangle Park, NC, USA

Hugh D.C. Smyth
College of Pharmacy
The University of Texas at Austin
Austin, TX, USA

ISSN 2522-834X   ISSN 2522-8358   (electronic)
AAPS Introductions in the Pharmaceutical Sciences
ISBN 978-3-030-42785-6   ISBN 978-3-030-42783-2   (eBook)
https://doi.org/10.1007/978-3-030-42783-2

This Springer imprint is published by the registered company Springer Nature Switzerland AG
The registered company address is: Gewerbestrasse 11, 6330 Cham, Switzerland

Preface

Interpreting physical, chemical, and biological phenomena as linear relationships between variables, or as simple functions of the variables, has been a significant scientific and mathematical strategy to elucidate these phenomena for centuries. It is often the case that the nature of linearity is to follow mathematical functions, for example, power, exponential, or logarithmic functions; nevertheless, the desire to fit data to simple predictable expressions is imbued in every scientist and engineer. From a philosophical standpoint, there is no reason to criticize this approach as it allows us to interpret the natural world and has a lofty heritage going back to the classical world.

However, nonlinear phenomena have been identified in many fields and interpreted as periodic, catastrophic, chaotic, or complex, and involve a variety of mathematical tools for analysis. Benoit Mandelbrot's now classic book on the fractal geometry of nature and the many subsequent texts, for example, Wolfram's *A New Kind of Science*, have raised questions about the nature of reality and the interpretation of observed phenomena. It seems clear that the complexity of dynamic events (on any scale) can rarely be explained by linear interpretations. The rare exceptions are likely to represent a convergence of multiple phenomena giving the appearance of a linear relationship between variables.

In fields related to pharmaceutical sciences, some texts have been written by pioneers such as Brian Kaye. *A Random Walk Through Fractal Dimensions* and *Chaos & Complexity* were seminal volumes for the editors. Tracing the mathematics of complexity back to the nineteenth century and beyond gives a validity to the search for more accurate interpretations of experimental observations that should impact the pharmaceutical sciences as significantly as other fields of endeavor.

Chemistry and physics literature is replete with papers on complexity from such notables as Ilya Prigogine and Murray Gell-Mann. A broad range of biological phenomena, the most complex imaginable from molecular biology to ecology, are now the subject of complexity analysis. Pharmaceutical sciences encompass the biology, chemistry, physics, and mathematics associated with drug discovery, delivery, disposition, and action. This text describes a range of topics of importance in the pharmaceutical sciences that indicate a need for a nonlinear interpretation if they are to be characterized accurately, understood fully, and potentially controlled or modulated in the service of improved therapeutic strategies.

It is likely that the future will involve increasingly complex interpretations of data related to drug design and delivery, particularly as our knowledge of the human genome leads inexorably to the potential for individualized therapy. We hope that this text will promote discussion of the varied phenomena leading to pharmacological effects and complex interactions that result in improved disease control and health maintenance.

This volume is revised and updated. The increasing use of machine learning and artificial intelligence to navigate large datasets and establish correlations that can be probed for causation may lead to rapid therapeutic lead identification, drug product development, and clinical application. Perhaps one of the most immediately exciting prospects from these developments is rapid drug repurposing.

Research Triangle Park, NC, USA   Anthony J. Hickey
Austin, TX, USA   Hugh D.C. Smyth
December 2019

Acknowledgment

The authors would like to thank Dr. Vinicius Alves for his input on machine learning and artificial intelligence in drug discovery. We would also like to thank Carolyn Spence for her support and encouragement in initiating the project. It was a pleasure working with Priyadharsini Mothilalnehru whose conscientious attention in guiding the production of the book made it effortless for us.

Contents

Chapter 1
The Nature of Complexity and Relevance to Pharmaceutical Sciences

General systems science discloses the existence of minimum sets of variable factors that uniquely govern each and every system. Lack of knowledge concerning all the factors and failure to include them in our integral imposes false conclusions. Let us not make the error of inadequacy in examining our most comprehensive inventory of experience and thoughts regarding the evoluting affairs of all humanity.
Buckminster Fuller, 1975
Synergetics, MacMillan Publishing Co, New York

Abstract The ability to manage large data sets in real time has been evolving over the last 50 years. Prior to the advent of rapid, high-capacity computing capabilities, the approach to data analysis had been to develop analytical approximations from which metadata could be derived. As the potential to draw direct inferences from data has increased, a new mathematics has developed in parallel that considers the complex functions that underpin natural phenomena. Pharmaceutical sciences are a microcosm of the more universal trends in science and society. Since the first edition of this book, it has become commonplace to consider the data to wisdom paradigm. In fact, there have been NIH and NSF initiatives that were intended to capitalize on this view of data treatment in the hopes of developing new taxonomies and ontologies. The topic of nonlinearity in physical and information phenomena is not new but is of increasing interest in developing rational strategies to address drug product development activities.

Keywords Non-linearity · Chaos · Complexity · Data · Information · Knowledge · Wisdom · Pharmaceutical Development

The exponential increase in data generated since the turn of the millennium and the ability to conduct rapid, computer-assisted searches have positioned the pharmaceutical sciences for disruptive change (Vamathevan et al., 2019). As further insight is gained and novel strategies emerge, it is worth reflecting on the complex phenomena

© American Association of Pharmaceutical Scientists 2020   1
A. J. Hickey, H. D.C. Smyth, *Pharmaco-complexity*, AAPS Introductions in the
Pharmaceutical Sciences, https://doi.org/10.1007/978-3-030-42783-2_1

that have challenged those in research and development to adopt new approaches to rationally address the urgent unmet needs in emerging diseases and for new product development.

The decade of the 1970s was a period of change for many reasons. In addition to the revolution in personal computing and information and gaming technology, a new view of the underpinning mathematics and science in nature was evolving. Benoit Mandelbrot developed a mathematical approach to describe the apparent complexity in nature (Mandelbrot, 1977). He observed that the superficial appearance of objects was built on a foundation of nonlinearity or "roughness" which could be described simply as self-similar at any scale of scrutiny. This "fractal" approach became a central philosophy behind complexity studies. As scientists in different disciplines began to accommodate complex interpretations of their data, a new approach to predicting or describing physical phenomena evolved that challenged more traditional methods and increased the potential to understand previously poorly understood. During this period others were also formulating approaches to nonlinear phenomena in a variety of fields (Woodcock & Davis, 1978). James Gleick's popular book *Chaos* describes in detail the exciting story of the birth and development of the field of complexity in this period and the skepticism with which the original protagonists were greeted by the mainstream scientific community (Gleick, 1987). Three decades after these developments, there is sufficient evidence for the nonlinearity of many natural phenomena to propose that we might begin to interpret similar observations in pharmaceutical sciences.

The major fields/topics within the pharmaceutical sciences are no less open to these new advances than any other area, but there has yet to be a treatment of these broad principles in a single text. This volume is intended to inform those who have an interest and to make accessible some of the underlying principles in the hopes that a new generation of researchers will gain insight into the complexity associated with pharmaceutical systems. The following brief overview attempts to identify important areas in which new mathematical approached may prove useful in future data interpretation.

A range of phenomena have been identified that have multiple underlying mechanisms that may lead to the appearance of simple or complex outcomes which have historically been reduced to best fit mathematical interpretations that do not necessarily result in greater understanding. The data usually encodes much more information about the phenomenon being studied than can be derived from a simple best fit model. The challenge is to differentiate the information from random events in order to create knowledge about the system that can be used to promote understanding with implications for control and quality of experiments or products. To guide our thinking on this subject, Fig. 1.1a has been used to indicate the foundation of data that creates information suitable for knowledge of a system and hopefully creating the wisdom to predict and control phenomena of importance. After a little consideration, it will occur to most scientists that we more frequently take a limited body of data or data from which information is being derived. Moreover, it is apparent knowledge and wisdom derived from this data is perhaps not to the extent that we would desire as shown in Fig. 1.1b. In both of these depictions, the conventional assumption is that inductive reasoning evolves the understanding required to manage the process or system under evaluation.

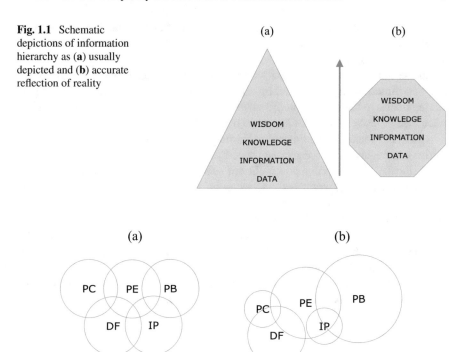

Fig. 1.1 Schematic depictions of information hierarchy as (**a**) usually depicted and (**b**) accurate reflection of reality

Fig. 1.2 Schematic diagrams of the relationship between elements of pharmaceutical development: (**a**) classical Venn diagram depicting perceived overlap/interaction; (**b**) extrapolation to capture magnitude of contribution of each. PC, physicochemical factors; DF, dosage form; PE, pharmaceutical engineering (manufacturing); IP, individual pharmacokinetics and pharmacodynamics; PB, population biology

The following text has been divided into major areas of importance in the pharmaceutical sciences which can be related to the product development road map: beginning in the area of drug discovery, which includes chemistry, pharmacology, and biochemistry of the drug molecule and its biological target (receptor, enzyme, antibody); extending through the dosage form considerations (solid-state chemistry and process engineering) into disposition following preclinical or clinical administration to animals or humans, respectively; and finally to the importance of population biology to the therapeutic effect. In addition, the advent of machine learning and artificial intelligence allows large data sets in each of the areas considered in the described to be explored rapidly and efficiently for drug discovery, formulation, product development, biological impact, and clinical outcome. Many of the tools used to interrogate the data are mechanism independent and simply derive inferences from large data sets that can be used to support hypotheses. These are example topics from within the range of items that must be considered in product development. Figure 1.2a illustrates the overlap between these fields. Figure 1.2b begins to accommodate some of the other elements of interactions such as the scale of the contribution (as indicated by the size of each circle). This is not intended to elicit discussion of the accuracy of the interpretation; rather it is to open up a discussion of whether we grasp the subtleties of the confounding factors and their contribution to our ultimate ability to understand, predict, and control important factors in product performance and therapeutic effect.

In the general conclusion, we will return to these themes and particularly the implicit directionality (indicated by arrows in the diagrams) of our thinking about our approach to the generation of knowledge and understanding of the influence of each discipline on others that appear to overlap.

Computer modeling and machine learning are being applied to complex data to derive ontologies that govern most aspects of the pharmaceutical product development. Ideally, they improve probability of success and as a consequence efficiency with respect to time, resources, and expense. Informed critical path management supports affordable, safe, and effective therapies and is likely to reshape the future of biomedical research and the pharmaceutical industry.

The reader is directed to further texts for general reading on topics that are broadly relevant to the subject of this book (Banchoff, 1990; Barnsley, 1985; Efros, 1986; Falconer, 1990; Kaye, 1993; Kim & Stringer, 1992; Liebovitch, 1998; Moon, 1992; Nayfeh & Balachandron, 1995; Schroeder, 1991; Stauffer & Aharony, 1992; Thompson & Bishop, 1994; Vamathevan et al., 2019).

References

Banchoff, T. (1990). *Beyond the third dimension*. New York, NY: Scientific American Library.

Barnsley, M. (1985). *Fractals everywhere*. New York, NY: Academic Press.

Efros, A. (1986). *Physics and geometry of disorder*. Moscow, Russia: Mir Publishers.

Falconer, K. (1990). *Fractal geometry, mathematical foundations and applications*. New York, NY: Wiley.

Gleick, J. (1987). *Chaos*. New York, NY: Penguin Books.

Kaye, B. (1993). *Chaos and complexity: Discovering the surprising patterns of science and technology*. New York, NY: VCH Publisher.

Kim, J., & Stringer, J. (1992). *Applied Chaos*. New York, NY: Wiley.

Liebovitch, L. (1998). *Fractals and Chaos, simplified for the life sciences*. Oxford, UK: Oxford University Press.

Mandelbrot, B.B. (1977). *The Fractal Geometry of Nature*, NY: WH Freeman.

Moon, F. (1992). *Chaotic and fractal dynamics*. New York, NY: Wiley.

Nayfeh, A., & Balachandron, B. (1995). *Applied nonlinear dynamics*. New York, NY: Wiley.

Schroeder, M. (1991). *Fractals, Chaos, power laws*. New York, NY: WH Freeman and Company.

Stauffer, D., & Aharony, A. (1992). *Introduction to percolation theory* (Second ed.). Washington, DC: Taylor and Francis.

Thompson, J., & Bishop, S. (1994). *Nonlinearity and Chaos in engineering dynamics*. New York, NY: Wiley.

Vamathevan, J., Clark, D., Czodrowski, P., Dunham, I., Ferran, E., Lee, G., … Zhao, S. (2019). Applications of machine learning in drug discovery and development. *Nature Reviews Drug Discovery, 18*(6), 463–477.

Woodcock, A., & Davis, M. (1978). *Catastrophe theory*. New York, NY: Penguin Books.

Chapter 2
Phenomena in Physical and Surface Chemistry

Abstract Complex phenomena leading to nonlinear behavior occur at all scales of scrutiny. The underlying complexity in chemical and physicochemical phenomena has been noted. Chemical reactions, molecular aggregation, and self-association have been shown to follow principles that require nonlinear interpretation to allow prediction of system behavior. Some phenomena enter the realms of mathematical complexity. Notably, gas adsorption isotherms/desorption isotherm can exhibit catastrophic events in which subtle changes in conditions lead to discontinuous changes in surface occupancy. Additionally, modeling of surface rugosity and porosity with classical analyses, such as Fourier series interpretations, can evolve into fractal non-integer dimensions that are between traditional surface and volume terms.

Keywords Physical and surface/interfacial chemistry · Reactions · Aggregation · Adsorption · Catastrophe · Solid surface

Complex phenomena in pharmaceutical chemistry may have a significant impact on the performance of the dosage form. Traditionally the concern has been with bulk product performance and stability with a focus on chemical degradation. This remains a serious issue, but the importance of accuracy and reproducibility of product performance as a metric of product quality has elevated component compatibility and physical stability to prominence.

A brief discussion of chemical considerations is followed by a more detailed presentation of the development of our understanding of important physical and surface chemical considerations.

2.1 Chemical Reactions

Certain chemical reactions are known to be complex phenomena. Molecular dynamics and chemical kinetics have been studied thoroughly and frequently appear predictable in terms of classical exponential or power functions (Billing & Mikkelsen, 1996). The apparently predictable nature of chemical reactions may be an illusion of simplicity described by the Ergodic hypothesis (Oliveira & Werlang, 2007; Szasz, 1994). Boltzmann originally conjectured that time-averaged behavior of micro-

© American Association of Pharmaceutical Scientists 2020
A. J. Hickey, H. D.C. Smyth, *Pharmaco-complexity*, AAPS Introductions in the Pharmaceutical Sciences, https://doi.org/10.1007/978-3-030-42783-2_2

5

scopic components of a system gives the same outcome as the macroscopic, bulk average where the bulk is a collection of all possible states that molecules would reach in assembly in infinite time. Clearly depicting each of these states independently would be an enormously complex system. Attempts are being made to scrutinize reactions to more accurately depict the events that may be occurring (Bonchev, Kamenski, & Temkiń, 1987; Nowak & Fic, 2010).

2.2   Surface and Interfacial Chemistry

Adsorption constitutes an important area of research in which efforts to understand complex phenomena have extended for over a century. The following section begins by considering molecular association and extends to surface adsorption and models that have been proposed for nonlinear data fitting. The ability to measure surface features and energy densities at a molecular level has improved the potential to learn about the nature of interactions.

Surfactant molecules are noteworthy because of their capacity through discrete polar and nonpolar regions to align at interfaces, in particular the surfaces of solids in suspension.

The use of surfactants as coating materials requires consideration of the nature of these compounds and their interactions and association with other substances.

Surface activity is a dynamic phenomenon, since the final state of a surface or interface represents a balance between the tendency toward adsorption and toward complete mixing due to the thermal motion of molecules.

A surface-active agent (surfactant) may be described as a substance which alters the conditions prevailing at the interface. All surfactants are characterized by two structural regions: a hydrocarbon chain, which is hydrophobic, and a polar, hydrophilic group. The nature of the hydrophilic region of the surfactant enables the classification of surfactants to be subdivided into anionic, cationic, and nonionic. Examples of these are sodium dodecyl sulfate, dodecyl trimethyl ammonium bromide, and n-dodecyl hexaoxyethylene glycol monoether, respectively. Two further groups of surfactants exist: ampholytic surfactants which are zwitterionic and can behave as any of the aforementioned examples depending on the pH at which they are maintained, such as alkyl betaine, and natural surfactants which usually contain a glycerol moiety, such as phosphatidylcholine. Therefore, surfactants may be described as amphiphilic.

The formation of aggregates (Kertes, 1977) and micelles (Eicke, 1977; Ravey, Buzier, & Picort, 1984) in solutions of surfactants is well documented. The term "micelle" should designate any soluble aggregate spontaneously and reversibly formed from amphiphilic molecules or ions (Tanford, 1980). The micellization processes according to the commonly used equilibrium thermodynamical descriptions, namely, the multiple equilibrium model and the pseudophase model, are, like the micelle definition, equally well applicable to aqueous and nonpolar solutions (Mukerjee, 1974). The second model best conforms to the definition of a micelle described above.

The interactions governing the formation of surfactant aggregates in apolar media are different from those in aqueous solutions, in spite of the apparently similar building principle of lipophilic and hydrophilic micelles. The differences between interactions encountered in aqueous and nonpolar surfactant solutions have been considered at a molecular level with reference to the stability or existence of micelles (Eicke, 1977). It has been concluded that once equilibrium between monomers and micelles, equivalent to the pseudophase model, ceases to be operative and is replaced by a stepwise aggregation equilibrium, the concept of a critical micelle concentration, CMC, is inapplicable. The two models for the process of interaction between surfactant molecules, described above, are, therefore, considered to be mutually exclusive (Kertes, 1977).

2.2.1 Aggregation

Inverted or reversed micelles are an example of molecular aggregation. Formation of surfactant aggregates has been referred to briefly above. The driving force for aggregation in aqueous media is the extrusion of hydrocarbon chains from solution upon micelle formation, resulting in an overall decrease in the free energy of the system. In nonaqueous solution, aggregation of the surfactant molecules depends upon both the solvent and surfactant structure.

Interactions between solvent and surfactant hydrocarbon chain groups tend to minimize the size of the aggregate, while interactions between the polar groups of the surfactants promote aggregation, in polar solvent.

Kinetic treatments in both aqueous and nonaqueous micellar systems have been based on the Hartley model (Hartley, 1936, 1955) of opposing hydrophobic interactions and electrostatic repulsions which are responsible for micellization in water.

Surfactant association in apolar solvents is predominantly the consequence of dipole-dipole and ion pair interactions between the amphiphiles. This differs from the Hartley model, and concepts derived for surfactant association in water may not necessarily be applicable to those in apolar solvents (O'Connor & Lomax, 1983).

In a nonaqueous solution of concentration, C, existing as simple molecules, m, and micelles composed of n molecules in equivalent concentration, M_n, the mass law is as follows:

$$KM_n = m^n = (C - nM_n)^n \qquad (2.1)$$

where K is the dissociation constant of the micelles.

A phase separation model was advocated by Shinoda and Hutchison (Shinoda & Hutchinson, 1962) and successfully applied by Singleterry (Singleterry, 1955) and Fowkes (Fowkes, 1962) to describe the aggregation of dinonylnaphthalene sulfonates in benzene. The phase separation model postulates that micellization is a phase transition. In its simplest form, it does not contain a size-limiting step, and, therefore, it is of little value in accounting for the formation of the small aggregates seen in apolar media.

The application of the mass action law to the overall aggregation process:

$nm \overset{K_n}{\leftrightarrow} M_n$, where K_n is the association constant of the process allows a model to be proposed. Thus, with the conservation of mass:

$$[m]/[D] + n([m]/[D])^n [D]^{n-1} K_n = 1 \tag{2.2}$$

where $[D]$ is the total molal concentration of detergent. From this equation assuming that K_n is the product of $n-1$ individual and equal mass action constants, the following equation is obtained:

$$[m]/[D] + n[M_n]/[D] = 1, \tag{2.3}$$

and the CMC is then obtained:

$$\text{CMC} = 1/k = [D] \tag{2.4}$$

where $[D] = 1$.

This system allows for two aggregational states, monomers and micelles. This does not account for the distribution in molecular weights. Smooth transitions from monomer, dimmer, trimer, etc. with concentration-dependent growth of aggregates have been observed. These gradual physical changes have been described (Lo et al., 1975) in terms of a sequential-type self-association model. Assuming all values of equilibrium concentration are the same, K_{12}, $K_{23} - K_{ij}$ are assumed to be equal:

$$\text{Monomer} + \text{Monomer} \overset{K_{12}}{\leftrightarrow} \text{Dimer}$$

$$\text{Dimer} + \text{Monomer} \overset{K_{23}}{\leftrightarrow} \text{Trimer}$$

$$(n-1)\text{mer} + \text{Monomer} \overset{K_{ij}}{\leftrightarrow} n-\text{mer}$$

Then the weight fraction of the monomer, f, is related to the stoichiometric concentration of detergent, $[D]$, by the following equation:

$$(1 - f^{1/2})/f = K_{ij}[D] \tag{2.5}$$

It would be possible to modify the multiple equilibrium model to account for a critical concentration (Eicke, 1980). Application of the mass action law to aggregation and conserving mass with respect to monomer yields

$$[m]/[D] + \sum_{n-2}^{n} n([m]/[D])^n [D]^{n-1} \prod_{n-2}^{n} k_{n-1} = 1 \tag{2.6}$$

A size-limiting step may be introduced by requiring a functional relationship between the equilibrium constants and the association number, $K_n = f(n)$ (Muto, Shimasaki, & Meguro, 1974).

The number of monomers involved in most surfactant aggregates in nonpolar solvents is relatively small (typically less than 10 for alkylammonium carboxylates compared with up to 100 for aqueous micelles (Fendler & Fendler, 1975); consequently, a spherical micelle structure would not provide effective shielding of the polar head groups from the solvent and its formation would be considered unlikely (Kertes & Gutmann, 1976). The alternative model is that of a lamellar micelle in which the polar and hydrophobic groups are placed end to end and tail to tail, with water and organic solvents between them (Mayer, Gutmann, & Kertes, 1969; Philippoff, 1950).

The kinetics of formation and decomposition of micelles and of the association-dissociation of the monomer to and from the micelle has rarely been studied in reversed micellar solutions (Yamashita, Yano, Harada, & Yasunaga, 1982). The paucity of data is attributed to (1) the aggregation number of the micelle being very low that an abrupt change in the physicochemical properties of the solution cannot be expected at the CMC and (2) micelle formation and monomer exchange reactions too rapid to be observed by conventional techniques.

2.2.2 Adsorption from Solution

The method by which surfactants interact with other substances is adsorption at the interface. The interface may be a gas-liquid, liquid-liquid, or solid-liquid juncture, the latter being significant in many pharmaceutical systems.

Langmuir presented a general equation for the isotherm of localized adsorption that was suitable for describing the adsorption of solutes. Langmuir's approach was concerned with monolayer coverage (Langmuir, 1917). The assumptions made for this model were that molecules are adsorbed at active centers on the adsorbent surface which they occupy for a finite period of time; owing to the small radius of action of adsorption forces and to their saturability, every active center while adsorbing molecules becomes incapable of further adsorption. Langmuir's adsorption equation concerns the simplifying assumptions that the heat of adsorption is independent of surface coverage, thus ignoring adsorbate interaction and the weakening effect on the intermolecular forces by distance between the adsorbent and adsorbate. Fowler and Guggenheim (Fowler & Guggenheim, 1960) adopted an approach which provides a modification for lateral interactions in the Langmuir model. Attempts have been made to generalize monolayer and multilayer concepts in order to describe the isotherms of different shapes by a single equation. Brunauer, Emmett, and Teller (Brunauer, Emmett, & Teller, 1938) developed such a generalized theory in respect to adsorption of vapors which has since become known as the BET theory. The assumptions of this theory are as follows: the adsorbent surface has a definite number of active sites which are equivalent energetically and are capable of retaining the adsorbate molecules; the interaction of the neighboring adsorbed molecules is neglected; the molecules in each layer act as an adsorption site for subsequently adsorbed molecules; and it is assumed that all of the adsorbed molecules in the second and subsequent layers have the same partition function as in the liquid state, which differs from the partition function of the first layer. Brunauer (Brunauer, 1945) has classified adsorption isotherms into five types.

The BET theory has been very useful in the interpretation of solute adsorption. This theory assumes that every molecule of a liquid has only two close neighbors, from the top and bottom chain, while the molecules of a real liquid are surrounded by many more adjacent molecules. Moreover, Giles (Giles, D'Silva, & Easton, 1974a, 1974b) has found both theoretically and experimentally the shape of some isotherms, of solute from solution, can be accounted for by postulating that adsorbate interaction does occur under particular conditions. The assumption that the adsorbent surface has a definite number of active sites which are equivalent energetically is an oversimplification (Mabire, Audebert, & Quivoron, 1984; Rudzinski & Narkiewicz-Michalek, 1982).

The Langmuir model for adsorption assumes that while the adsorbed molecules occupy sites of energy Q that they do not interact with each other. Fowler and Guggenheim (Fowler & Guggenheim, 1960) have adopted an approach which accounts for lateral interaction.

The probability of a given site of energy Q being occupied is N/S, and if each site has z neighbors, the probability of neighbor site being occupied is zN/S. Thus, the fraction of adsorbed molecules is $z\theta/2$, the factor one half correcting for double counting and θ being total monolayer coverage. If the lateral interaction energy is ω, the added energy of adsorption is $z\omega\theta/2$, and the added differential energy of adsorption is just $z\omega\theta$.

The modified Langmuir equation then becomes

$$\Theta = b'C\left(1+b'C\right)^{-1}$$

$$b' = b_0 \exp\left(Q+z\omega\theta\right)/\text{RT} = b\exp\left(z\omega\theta/\text{RT}\right) \qquad (2.8)$$

Rearranging these equations

$$bC = \theta/\left(1-\theta\right)\exp\left(-z\omega\theta/\text{RT}\right) \qquad (2.9)$$

This is equivalent to the Frumkin or Volmer (Damaskin, Petril, & Batrakov, 1971) expression:

$$bC = \theta/\left(1-\theta\right)\exp\left(-2a\theta\right) \qquad (2.10)$$

Therefore, $a = z\omega/2\text{RT}$, where a is the interaction energy between adsorbed molecules.

When lateral interactions are added to the Langmuir expression, the adsorption sigmoid isotherm contracts to a step function, and when the interaction energy between adsorbed molecules is exceeded ($2a = \beta = 4$), the adsorption isotherm takes on a mathematical catastrophic sigmoid appearance when θ is plotted against bC (refer to Fig. 2.1). The assumption is that the sigmoid function collapses to the dotted line as a two-phase equilibrium occurs and the gas condenses to a liquid. If this is the case, the logical next state is to consider equations of state that consider covolume and equation of state. Details of this approach are described by Adamson and involve van der Waals equations of state (Adamson, 1967). The constants in the van der Waals equation depend on volume and temperature. The complex adsorption equations that emerge from this approach. Figure 2.2 also depicts the relationship between θ and kC from van der Waals equation, where c is equivalent to β with an addition term in the denominator (σ^0) representing volume ($c = z\omega/\sigma^0 RT$). The van der Waals adsorption isotherms have been overlaid on the Langmuir lateral interaction energy isotherms in Fig. 2.1 for convenience. In reality the van der Waals equation of state isotherms follows approximately the same pattern as Langmuir but plateaus at higher kC (slightly further right on the figure) than Langmuir does with respect to bC. In confirming the contraction and ultimate catastrophic sigmoid event (van der Waals $c > 6$, Langmuir $\beta > 4$), the equations of state isotherms also support an interpretation that the dotted line in the figure represents two-dimensional condensation of the gas into a liquid.

The adsorption of organic substances may yield sigmoid or logarithmic isotherms (Giles et al., 1974a, 1974b), depending on the interaction between the adsorbed particles being predominantly attractive or repulsive.

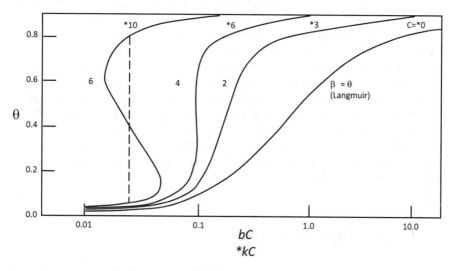

Fig. 2.1 Langmuir plus lateral interaction isotherms and a depiction∗ of the van der Waals equation of state isotherm. (Modified from (Adamson, 1967))

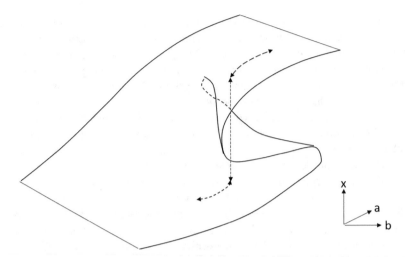

Fig. 2.2 The geometry of a two-dimensional cusp-catastrophe manifold where $F(x: a, b) = \frac{1}{4} x^4 + \frac{1}{2} ax^2 + bx$. The arrowed dotted line shows that a small change in the control parameters may lead to a sudden change in the state variable even if no catastrophe convention is followed. (Modified from Gilmore (1993a, 1993c))

The forces responsible for solute adsorption may be chemical or physicochemical and physical or mechanical (Giles, 1982). The chemical or physicochemical forces may be listed as covalent bonding, hydrogen bonds and other polar forces, ion exchange attraction' van der Waals forces, and hydrophobic forces. The physical or mechanical forces are restriction of movement of solute aggregates in micropores and facilitation of entry of solute by the progressive breakdown of the substrate structure. Under these circumstances rather than the event described for gases above where the sigmoid isotherm, shown in Fig. 2.1 representing high lateral interaction energies, simply collapses into the dotted line step function, the experimental measurements show the sigmoid function (Hickey, Jackson, & Fildes, 1988). The clear implication of the behavior of gases and solutes in solution is that there are other factors involved that are necessarily quantified in the experiment, and this can be illustrated from cusp geometry in catastrophe theory as shown in Fig. 2.2 (Gilmore, 1993c). Indeed, the collapse to the dotted line event shown in Fig. 2.2 is predicted by catastrophe theory where small changes in the control value (independent variable) will give rise to a sudden (discontinuous) change in the state variable equivalent to the condensation described by Langmuir and van der Waals isotherms (Gilmore, 1993a, 1993b).

2.2.3 Solid Surface Interaction

Figure 2.3a illustrates the classical model of molecules associating at the surface of particles. However, it has long been known that surfaces exhibit roughness and asperities that contribute fundamentally to interactions. Figure 2.1b illustrates a

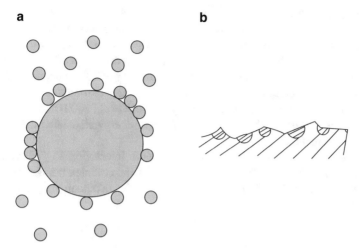

Fig. 2.3 (a) Classical model of surface adsorption phenomenon (small circles represent gas or solute molecules impinging at the surface of an idealized, spherical particle) and (b) schematic of particle surface depicting surface asperities (irregularity) and high energy density sites (indicated by hatched hemispherical areas)

more realistic depiction of the surface of a crystalline particle with areas of higher energy density (Hickey et al., 2007). The potential sources of these higher energy density sites may be the presence of: amorphous material, moisture, impurities, electrical charge and sites for mechanical interlocking. It should be evident from these images that depending on the scale of the means of measuring the surface a realistic surface may seem bigger or smaller based on the ability to penetrate into the small invaginations in the surface. Driven by knowledge of the true nature of surfaces, morphological approaches have included Fourier analysis and fractal analysis to approximate surfaces (Beddow, Vetter, & Sisson, 1976; Kaye, 1993; Meloy, 1977). Briefly Fourier analysis involves describing a particle surface by assigning a center to the image of the particle and then using polar coordinates to plot variations of the surface on a linear scale (Luerkins, 1991). This image can then be subjected to harmonic analysis from which a Fourier series and the respective coordinates each corresponding with a shape, sphericity, triangularity, etc. can be derived to describe the particle based on its surface. The principle of fractal geometry is that when periphery of an object, such as a particle, is measured at different scales of scrutiny, with yardsticks of different length, it appears to be larger as the scale is smaller (Kaye, 1989). There is a linear relationship between the exponent of the estimate of periphery and that of the dimension of the scale being employed. From the slope of this line, the so-called fractal dimension can be derived. This is often thought of as representing dimensions between 1 and 2 for flat, 2D images and between 2 and 3 for 3D images.

Beyond the morphology of individual particles, powders, composed of numerous particles, have been probed with molecules, each with a radius that establishes the scale of scrutiny to derive fractal dimensions (Avnir, 1989; Avnir & Farin, 1983;

Avnir, Farin, & Pfiefer, 1984; Pfeifer & Avnir, 1983). These experiments which employ inert gas molecules of different dimensions have been extended to consider the functional status of the surface. It could be argued that the use of the Scatchard interpretation of a Langmuirian adsorption but based on the specific binding capacity of proteins was an early example of a functional surface interaction (Xu et al., 2010). However, techniques such as inverse-phase gas chromatography (Telko & Hickey, 2007) and atomic force microscopy (Danesh et al., 2000) allow direct probing of surface features with respect to the force or energy associated with potential molecular interaction.

The recent extension of this molecule-particle interaction to particle-particle interactions may lead to greater understanding of the complex interactions of particles in pharmaceutical formulation which will be taken up in great detail in Sect. 2.3.

2.3 Summary

Chemical reactions are typified by apparently predictable behavior which masks a complex range of underpinning states. Interest in this subject is beginning to emerge in chemistry and may ultimately be useful in considering subtleties of chemical stability of pharmaceuticals (Carstensen, 1990). The nature of surfaces and their potential to interact with molecules or particles stems from a long and extensive history of attempts to model nonlinear phenomena in surface and interfacial chemistry. The strong foundation in this field gives a major opportunity for new developments as complex interpretations are employed to elucidate surface phenomena. The implications of greater understanding in this area with respect to important pharmaceutical phenomena such as deaggregation, dissolution, diffusion, and ultimately drug availability are taken up in the next section.

References

Adamson, A. (1967). Adsorption of gases and vapors on solids. In *Physical chemistry of surfaces* (2nd ed., pp. 565–648). New York, NY: Wiley.

Avnir, D. (1989). *The fractal approach to heterogeneous chemistry, surfaces, colloids, polymers.* New York, NY: John Wiley and Sons.

Avnir, D., & Farin, D. (1983). Chemistry in noninteger dimensions between two and three. II fractal surfaces of adsorbents. *The Journal of Physical Chemistry, 79*, 3566–3571.

Avnir, D., Farin, D., & Pfiefer, P. (1984). Molecular fractal surfaces. *Nature, 308*, 261–263.

Beddow, J., Vetter, A., & Sisson, K. (1976). Powder metallurgy review 9, Part I, particle shape analysis. *Powder Metallurgy International, 8*, 69–76.

Billing, G., & Mikkelsen, K. (1996). *Introduction to molecular dynamics and chemical kinetics.* New York, NY: Wiley.

Bonchev, D., Kamenski, D., & Temkin, O. (1987). Complexity index for linear mechanisms of chemical reactions. *Journal of Mathematical Chemistry, 1*, 345–388.

Brunauer, S. (1945). *The adsorption of gases and vapors I*. Princeton, NJ: Princeton University Press.

Brunauer, S., Emmett, P., & Teller, E. (1938). Adsorption of gases in multimolecular layers. *Journal of the American Chemical Society, 6*, 309–319.

Carstensen, J. (1990). *Drug stability principles and practices*. New York, NY: Marcel Dekker.

Damaskin, B., Petril, O., & Batrakov, V. (1971). *Adsorption of organic compounds on electrodes*. New York, NY: Plenum Press.

Danesh, A., Chen, X., Davies, M., Roberts, C., Sanders, G., Tendler, S., ... Wilkins, M. (2000). The discrimination of drug polymorphic forms from single crystals using atomic force microscopy. *Pharmaceutical Research, 17*, 887–890.

Eicke, H. (1977). Micelle in apolar media. In K. Mittal (Ed.), *Micellization, solubilization and microemulsions*. New York, NY: Plenum Press.

Eicke, H. (1980). Aggregation in surfactant solutions: Formation and properties of micelles and microemulsions. *Pure and Applied Chemistry, 52*, 1349–1357.

Fendler, J., & Fendler, E. (1975). *Catalysis in micellar and macromolecular systems*. New York, NY: Academic Press.

Fowkes, F. (1962). The micelle phase of calcium dinonylnaphthalene sulfonate in n-decane. *The Journal of Physical Chemistry, 66*, 1843–1845.

Fowler, R., & Guggenheim, E. (1960). *Statistical thermodynamics*. Cambridge, UK: Cambridge University Press.

Giles, C. (1982). Forces operating in adsorption of surfactants and other solutes at solid surfaces: A survey. In K. Mittal & E. Fendler (Eds.), *Solution behavior of surfactants. Theoretical and applied aspects*. New York, NY: Plenum Press.

Giles, C., D'Silva, A., & Easton, I. (1974a). A general treatment and classification of the solute adsorption isotherm I. Theoretical. *Journal of Colloid and Interface Science, 47*, 755–765.

Giles, C., D'Silva, A., & Easton, I. (1974b). A general treatment and classification of the solute adsorption isotherm II. Experimental interpretation. *Journal of Colloid and Interface Science, 47*, 766–778.

Gilmore, R. (1993a). Catastrophe flags. In *Catastrophe theory for scientists and engineers* (pp. 158–183). Mineola, NY: Dover Publications, Inc (Originally John Wiley and Sons, NY, 1981).

Gilmore, R. (1993b). Fluid equation of state: van der Waals equation. *In Catastrophe Theory for Scientists and Engineers* (pp. 203–210). Mineola, NY: Dover Publications, Inc. (originally John Wiley and Sons, NY, 1981).

Gilmore, R. (1993c). Geometry of the fold and cusp. In *Catastrophe theory for scientists and engineers* (pp. 94–106). Mineola, NY (Originally John Wiley and Sons, NY, 1981): Dover Publications, Inc.

Hartley, G. (1936). *Aqueous solutions of paraffin chain salts*. Paris, France: Heman and Cie.

Hartley, G. (1955). *Progress in chemistry of fats and other lipids*. London, UK: Pergamon Press.

Hickey, A., Jackson, G., & Fildes, F. (1988). Preparation and characterization of disodium fluorescein powder in association with lauric and capric acids. *Journal of Pharmaceutical Sciences, 77*(9), 804–809.

Hickey, A., Mansour, H., Telko, M., Xu, Z., Smyth, H., Mulder, T., ... Papadopoulos, D. (2007). Physical characterization of component particles included in dry powder inhalers. II. Dynamic characteristics. *Journal of Pharmaceutical Sciences, 96*, 1302–1319.

Kaye, B. (1989). *A random walk through fractal dimensions*. New York, NY: VCH Publishers.

Kaye, B. (1993). Applied fractal geometry and the fineparticle specialists. Part I. Rugged boundaries and rough surfaces. *Particle and Particle Systems Characterization, 10*, 99–110.

Kertes, A. (1977). Aggregation of surfactants in hydrocarbons, incompatibility of the critical micelle concentration concept with experimental data. In K. Mittal (Ed.), *Micellization, solubilization and microemulsions*. New York, NY: Plenum Press.

Kertes, A., & Gutmann, H. (1976). Surfactants in organic solvents: The physical chemistry of aggregation and micellization. In E. Matijevic (Ed.), *Surface and colloid science* (Vol. Vol. 8, p. 193). New York, NY: Wiley.

Langmuir, I. (1917). The constitution and fundamental properties of solids and liquids II. Liquids. *Journal of the American Chemical Society, 39*, 1848–1906.

Lo, F., Escott, B., Fendler, E., Adams, E., Larson, R., & Smith, P. (1975). Temperature-dependent self-association of dodecylammonium propionate in benzene cyclohexane. *The Journal of Physical Chemistry, 79*, 2609–2621.

Luerkins, D. (1991). *Theory and applications of morphological analysis, fine particles and surfaces.* Boca Raton, FL: CRC Press.

Mabire, F., Audebert, R., & Quivoron, C. (1984). Flocculation properties of some water soluble cationic copolymers towards silica suspensions: A semiquantitative interpretation of the role of molecular weight and cationicity through a "patchwork" model. *Journal of Colloid and Interface Science, 97*, 122–136.

Mayer, I., Gutmann, H., & Kertes, A. (1969). In A. Kertes & Y. Marcus (Eds.), *Solvent extraction research.* New York, NY: Wiley-Interscience.

Meloy, T. (1977). Fast Fourier transforms applied to shape analysis of particle silhouettes to obtain morphological data. *Powder Tech, 17*, 27–35.

Mukerjee, P. (1974). Micellar properties of drugs: Micellar and non-micellar patterns of self association of hydrophobic solutes of different molecular structures, monomer fraction, availability, and misuses of micellar hypotheses. *Journal of Pharmaceutical Sciences, 63*, 972–981.

Muto, S., Shimasaki, Y., & Meguro, K. (1974). The effect of counterion on critical micelle concentration of surfactant in non-aqueous media. *Journal of Colloid and Interface Science, 49*, 173–176.

Nowak, G., & Fic, G. (2010). Search for complexity generating chemical transformations by combining connectivity analysis and cascade transformation patterns. *Journal of Chemical Information and Modeling, 50*, 1369–1377.

O'Connor, C., & Lomax, T. (1983). Evidence for the sequential self-association model in reversed micelles. *Tetrahedron Letters, 24*, 2917–2920.

Oliveira, C. D., & Werlang, T. (2007). Ergodic hypothesis in classical statistical mechanics. *Revista Brasileira de Ensimo de Fisica, 29*, 189–201.

Pfeifer, P., & Avnir, D. (1983). Chemistry in noninteger dimensions between two and three. I. Fractal theory of heterogeneous surfaces. *The Journal of Physical Chemistry, 79*, 3558–3565.

Philippoff, W. (1950). Micelles and X-rays. *Journal of Colloid Science, 5*, 169–191.

Ravey, J., Buzier, M., & Picort, C. (1984). Micellar structures of non-ionic surfactants in apolar media. *Journal of Colloid and Interface Science, 97*, 9–25.

Rudzinski, W., & Narkiewicz-Michalek, J. (1982). Adsorption from solutions onto solid surfaces. *Journal of the Chemical Society Faraday Transactions I, 78*, 2361–2368.

Shinoda, K., & Hutchinson, E. (1962). Pseudo-phase separation model for thermodynamic calculations on micellar solutions. *The Journal of Physical Chemistry, 66*, 577–582.

Singleterry, C. (1955). Micelle formation and solubilization in non-aqueous solvents. *Journal of the American Oil Chemists' Society, 32*, 446–452.

Szasz, D. (1994). *Boltzmann's ergodic hypothesis, a conjecture for centuries?* Wien, Austria: The Erwin Schrodinger International Institute for Mathematical Physics.

Tanford, C. (1980). *The hydrophobic effect* (2nd ed.). New York, NY: Wiley.

Telko, M., & Hickey, A. (2007). Critical assessment of inverse gas chromatography as means of assessing surface free energy and acid-base interactions of pharmaceutical powders. *Journal of Pharmaceutical Sciences, 96*, 2647–2654.

Xu, Z., Mansour, H., Mulder, T., McLean, R., Langridge, J., & Hickey, A. (2010). Heterogeneous particle deaggregation and its implication for therapeutic aerosol performance. *Journal of Pharmaceutical Sciences, 96*, 2647–2654.

Yamashita, T., Yano, H., Harada, S., & Yasunaga, T. (1982). Kinetic studies of the micelle system of octylammonium alkanoates in hexane solution by the ultrasonic absorption. *Bulletin of the Chemical Society of Japan, 55*, 3403–3406.

Chapter 3
Solid-State Pharmaceuticals: Solving Complex Problems in Preformulation and Formulation

Abstract Pharmaceutical systems and products almost always include solid-state ingredients either during manufacture or as the dosage form itself. Thus preformulation and formulation of drug products is often critically dependent on the characterization and understanding of these physicochemical properties. Despite their seemingly simplicity, attributes like dissolution, particle shape, and powder flow can be exceedingly complex and often require nonlinear approaches for modeling, prediction, and analysis. This section provides examples both from static and dynamic processes encountered in pharmaceutical preformulation/formulation development.

Keywords Physicochemical properties · Dynamical systems · Solubility · Dissolution · Fractal · Monte Carlo · Powder flow · Powder mixing

The preformulation and formulation of new chemical entities has evolved considerably in the past 50 years, being driven largely by dramatic changes in the physicochemical properties of newly discovered drugs and the greater complexities of regulation of these drug products. The objective of this chapter is to provide illustrative examples of several well-known issues in preformulation/formulation that have benefited from a nonlinear data analysis and interpretation. These examples have proved extremely challenging for scientists as they have attempted to develop drugs and drug products due to the inability of standard modeling or traditional mathematical approaches to mimic the processes or provide a predictive way forward. However, interpretation of the problem using different approaches originating from the fields of dynamical systems, chaos, complexity, fractals, tenable solutions, and an even deeper understanding has been facilitated. The chapter is divided into two sections: "Statics" and "Dynamics." Statics relates to the static properties of new chemical entities (NCEs) that are generally considered to be physical and chemical properties that are determined during preformulation. When the NCEs enter formulation stages of development, one has to consider more dynamic relationships of the drug and delivery system. This important distinction between static and dynamic systems with respect to complexity and data interpretation will allow readers to not only consider these presented issues but also extend these approaches to analysis to their own specific data set.

A. J. Hickey, H. D. C. Smyth, *Pharmaco-complexity*, AAPS Introductions in the Pharmaceutical Sciences, https://doi.org/10.1007/978-3-030-42783-2_3

3.1   Statics

Following lead optimization, preformulation studies are critical for candidate selection. This thorough characterization of the physicochemical properties of drugs is the foundation for developing robust dosage forms and ensuring development times and costs are optimized. Most developers of new drugs will perform preformulation studies that address several key physicochemical properties that are known to have a significant impact on the dynamics of the formulation (e.g., dissolution, biopharmaceutics, manufacturing, etc.). Characteristics of drugs determined during preformulation studies will give some indication of the likelihood of "drugability" and the optimal pathway for formulation success (e.g., Lipinski's rule of five) (Keller, Pichota, & Yin, 2006; Lipinski, Lombardo, Dominy, & Feeney, 2001). The focus of this section is to highlight some aspects of the physicochemical properties of drugs in which clear known examples of complexity have been established and briefly discuss the approaches taken by scientists to cope with this behavior.

3.1.1   Solubility, Dissolution, and Drug Release

Prediction of solubility, dissolution, and drug release is of great interest in the pharmaceutical industry. Solubility, for example, is now widely recognized as an important issue limiting and/or preventing drug development due to biopharmaceutical constraints preventing drug absorption at levels needed for therapeutic effect. Equilibrium solubility can be defined as the drug concentration in the molecular state in a solution, where the solvent is in contact with an excess amount of the solid compound and where the concentrations do not change over time. Solubility is key for most pharmaceutical systems that rely on absorption through molecular mechanisms. It is well recognized that drugs administered orally for systemic effect must have some degree of solubility in gastrointestinal fluids for the drug to be absorbed and exert a therapeutic effect. It is estimated that around 80–90% of NCEs are classified as having poor water solubility and may prevent or limit development of these potential therapies.

In 1897, Noyes and Whitney noticed that the rate of dissolution is proportional to the difference between the instantaneous concentration c at time t and the saturation solubility, c_S (Dokoumetzidis & Macheras, 2006).

$$\frac{dM}{dt} = \frac{DA}{h}\left(c_s - c_t\right) \tag{3.1}$$

In the pharmaceutical field, dissolution was once established to be an important issue for drug absorption and bioavailability, and many aspects of this process were investigated. Solubility, stirring, particle size, and wettability were among the key aspects studied by the rapidly expanding field of dissolution science. Initially these studies focused on the changes that could be induced in dissolution processes by modifying the environment and formulations. Subsequent emphasis has been on dissolution as a predictor of oral drug absorption, and this leads to a landmark paper introducing the Biopharmaceutics Classification System (BCS) by Amidon, Lennernas, Shah, and Crison (1995) (Amidon et al., 1995). Many studies then were published introducing new models that attempted to correlate in vitro studies to in vivo observations (e.g., Dressman, Amidon, Reppas, & Shah, 1998; Persson et al., 2005). Other routes of administration are also significantly influenced by solubility and dissolution limitations (Warnken, Smyth, & Williams, 2016). Despite advances in dissolution testing provided by these and other studies, the field has yet to overcome obstacles to modeling the in vivo situation and predicting drug solubility in various relevant media or pharmaceutical solvents. Some specific obstacles in this field include the complexity and chaotic nature of the hydrodynamic conditions of dissolution in vivo (D'Arcy, Corrigan, & Healy, 2005; D'Arcy, Corrigan, & Healy, 2006), the complexity of gastrointestinal drug absorption phenomena (Macheras & Argyrakis, 1997), and the heterogeneity of in vivo conditions. Excellent reviews of dissolution can be found in the recent literature (Dokoumetzidis & Macheras, 2006; Siepmann & Siepmann, 2013). Establishing models and tests for predicting solubility and dissolution through different methods is clearly needed.

3.1.2 Fractal Dimensions and Surface Phenomena

An introduction to fractals and fractal geometry is provided elsewhere in this text and in excellent texts on the subject (Barnsley & Rising, 1993; Kaye, 1994). Examples are shown in Fig. 3.1 and basic definitions given in Table 3.1. The application of fractals to surfaces has been largely facilitated by the work of Avnir and colleagues. They explored fractal geometry in chemistry and physics,

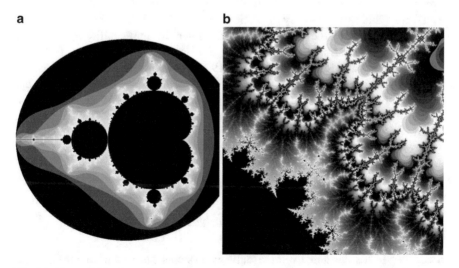

Fig. 3.1 Example of a fractal (Mandelbrot set) viewed at a distance (**a**) and at much higher magnification (**b**) indicating the appearance of self-similarity at these different scales. (Images produced on FractalWorks software)

Table 3.1 Fractal geometry primer

Definition	A fractal is an object/quantity that displays self-similarity across many scales of scrutiny. The object need not exhibit exactly the same structure at all scales, but the same "type" of structures must appear on all scales. A plot of the quantity on a log-log graph versus scale gives a straight line, whose slope is said to be the fractal dimension (Weisstein, 2010)		
Example	The length of a coastline measured with different length rulers. The shorter the ruler, the longer the length measured, a paradox known as the coastline paradox		
Examples in pharmaceutical sciences	Dissolution (Akbarieh & Tawashi, 1989; Farin & Avnir, 1992; Tromelin, Gnanou, Andrès, Pourcelot, & Chaillot, 1996; Tromelin, Hautbout, & Pourcelot, 2001)	Drug distribution, pharmacokinetics, and toxicity (Fuite, Marsh, & Tuszyński, 2002; Karalis & Macheras, 2002; Macheras, 1996; Marsh & Riauka, 2007; Marsh & Tuszynski, 2006; Pang, Weiss, & Macheras, 2007; Pereira, 2010)	Particle analysis (Avnir et al., 1991; Concessio & Hickey, 1997; Fini, Holgado, Rodriguez, & Cavallari, 2002)
	Imaging (Martin-Landrove, Pereira, Caldeira, Itriago, & Juliac, 2007; Zook & Iftekharuddin, 2005)	Drug diffusion (Liu & Nie, 2001; Roncaglia, Mannella, & Grigolini, 1994)	Pharmacodynamic responses (Jartti, Kuusela, Kaila, Tahvanainen, & Välimäki, 1998)

applying this mathematical approach to surface interactions (Avnir & Farin, 1984). Avnir and Farin showed at the molecular level the surfaces of most materials were fractal. They extended their work in fractal chemistry to drug dissolution. Notably, they provided a modified Noyes-Whitney equation and a Hixon-Crowell cube root law to include surface roughness effects on the dissolution rate of drugs (Farin & Avnir, 1992). Carstensen and Franchini (1993), however, argue that self-similarity of particles during dissolution may not be a valid assumption. However, several authors have shown the applicability of fractals to dissolution of pharmaceutical systems, for example, diclofenac (Holgado, Fernández-Hervás, Rabasco, & Fini, 1995), sodium cholate (Fini, Fazio, Fernández-Hervás, Holgado, & Rabasco, 1996), orthoboric acid (Tromelin et al., 1996), and others (Akbarieh, Dubuc, & Tawashi, 1987; Akbarieh & Tawashi, 1989; Farin & Avnir, 1992; Johns & Gladden, 2000; Schroder & Kleinebudde, 1995; Weidler, Degovics, & Laggner, 1998). These models have been extended to better predict solubility of nanoparticles compared to the Ostwald-Freundlich equation (Mihranyan & Stromme, 2007).

3.1.3 Monte Carlo Methods in Modeling Drug Release

Since the 1970s sustained release delivery systems have evolved rapidly. The release of drug from these systems must be predictable for the delivery system to be successful. The kinetics of drug release follows the operative release mechanism of the system, e.g., diffusion through inert matrix, diffusion across membrane or hydrophilic gel, osmotic-driven pump, ion exchange, etc. By far, diffusion is the most common release mechanism. Solute release models were described early, and modeling drug release from diffusion-controlled systems relies on the Higuchi model published in 1961 (Higuchi, 1961). The kinetics of release from an ointment was described, assuming homogeneously mixed drug in a planar matrix diffusing into a medium, a perfect sink, under pseudo steady-state conditions. This model has been widely used and was instrumental for the development and understanding of dosage form design. However, due to the approximate nature of the model, its use for the analysis of release data is recommended only for the first 60% of the release curve, beyond which, the model is insufficient. In 1985 Peppas (1985) introduced a semiempirical equation (the power law) to describe drug release from polymeric devices (Peppas, 1985; Siepmann & Peppas, 2001). Similarly, valid estimates of drug release can be obtained by fitting this equation to the first 60% of the experimental release data. This highly cited work has been widely applied in the analysis of drug release studies. Mechanistic release models have been published in literature (Siepmann & Peppas, 2001) and are more physically realistic, but their mathematical complexity is their main disadvantage for wide use.

More recently, methodology used to find the release mechanisms for an entire set of data has been demonstrated via the use of Monte Carlo simulations. Kosmidis and coworkers (Kosmidis & Macheras, 2007; Kosmidis & Macheras, 2008; Kosmidis, Rinaki, Argyrakis, & Macheras, 2003) showed that the Weibull function could be used to describe release kinetics in either Euclidean or fractal spaces. Based on these findings, a new methodology was developed for the description and determination of release mechanisms using the *entire* set of data (Papadopoulou, Kosmidis, Vlachou, & Macheras, 2006) (Table 3.2).

Table 3.2 Monte Carlo method

Definition	The use of randomly generated or sampled data and computer simulations to obtain approximate solutions to complex mathematical and statistical problems. Monte Carlo methods enable the investigation of complex and dynamical systems by direct study of their properties using computer simulation		
Examples in pharmaceutical sciences	Drug release (Kosmidis, Argyrakis, & Macheras, 2003; Kosmidis & Macheras, 2008)	Colloidal dispersions (Aoshima & Satoh, 2005; Narambuena, Ausar, Bianco, Beltramo, & Leiva, 2005; Sanz & Marenduzzo, 2010; Sjoberg & Mortensen, 1997; Wang, Sheng, & Tsao, 2009)	Patient compliance (Ahmad, Douglas Boudinot, Barr, Reed, & Garnett, 2005)
	Solubility prediction (Jorgensen & Duffy, 2000)	Pharmacokinetics (Dokoumetzidis, Kosmidis, Argyrakis, & Macheras, 2005; Montgomery et al., 2001)	Manufacturing and processing (Kuu & Chilamkurti, 2003; Rowe, York, Colbourn, & Roskilly, 2005)

3.1.4   Surfaces and Particles

Complexity of Crystallization

Over 90% of all pharmaceutical products contain active ingredients produced in crystalline form. Crystallization processes can illustrate some interesting dynamical behavior, including a high sensitivity to parameter variations. This can be a cause of considerable issues in production and also a significant impact on the efficiency and profitability of the overall crystallization process. Differences and variability in the production of the crystals can impact downstream processes too (often in a nonlinear and complex way): filtration, drying, milling, powder flow, surface energetics, bioavailability, tablet stability, etc. The dynamics of both continuous crystallization and batch crystallization processes have been investigated and reviewed (Braatz & Hasebe, 2002; Rawlings, Miller, & Witkowski, 1993). Crystallization processes are highly nonlinear and are modeled by coupled nonlinear algebraic integro-partial differential equations. However, due to large differences and length scales (Angstrom to micron) and time scales (microseconds to minutes), these equations are found not to be useful for nonlinear feedback controllers using available computers (Christofides, 2002). Therefore use of tools from nonlinear dynamics and complexity have been described for this problem and have been summarized in a recent review by Fujiwara and coworkers (Fujiwara, Nagy, Chew, & Braatz, 2005).

Complexity of Particle Shape

Particle morphology and particle size distributions can have profound effects on the manufacture, processing, and performance of pharmaceutical dosage forms. The shape of particles within a powder is rarely regular, but it is very valuable to be able to describe particle shape using quantitative measures. This can enable understanding of dynamics of drug dissolution, aerodynamics, and other important properties conferred by shape in pharmaceutical systems. It is also necessary to have a quantitative assessment of shape for quality control and regulatory purposes. As such, a variety of shape factors have been described in the literature (Concessio & Hickey, 1997). Static shape factors are those that can be obtained from imaging of particles, and they attempt to describe either the deviation from ideal shape (e.g., sphere) or shape independent of a physical reference. Most useful and widely described in modern literature is fractal analysis of particle shape. Fractal geometry and fractal surfaces are based on self-similarity of a surface, or features, at different scales. As one looks at an object at increasing or decreasing levels of scrutiny, the features and patterns are observed to be repetitive. Rugged structures such as powder particles can be assessed using fractal geometry via estimation of a mass fractal dimension or boundary fractal dimension (Kaye, 1978). Different methods have attempted to characterize particle shape using polar coordinates from which Fourier coefficients are derived (Luerkens, 1991). However, these methods are tedious, and use of surface area determinations across narrow particle size distributions has been suggested to be an alternative method of obtaining a fractal dimension estimate of particle shape (Concessio & Hickey, 1997).

3.2 Dynamics: Examples and Approaches of Handling Complexity in Formulation of Pharmaceuticals

This section focuses on solids and powders and their processing and formulation as an illustration of how dynamic complexity also pervades pharmaceutical development. Clearly dynamics in other areas of pharmaco-complexity exist and may be able to be interpreted using similar approaches.

3.2.1 Powder Flow and Mixing

The dynamic behavior of powders and granular matter are well known to be examples of collective systems that are far from equilibrium (Muzzio et al., 2003; Ottino & Khakhar, 2002). They have been used to illustrate nonlinear dynamics and complex systems as systems that experience self-organization, invariance and symmetry breaking, pattern formation (waves, chaos) (Gilchrist & Ottino, 2003; Hill, Khakhar, Gilchrist, McCarthy, & Ottino, 1999; Khakhar, McCarthy, Gilchrist, & Ottino, 1999; Ottino & Khakhar, 2002). Moreover, these concepts have also been applied across a wide range of scales from pharmaceutical fine particles to large-scale geological movement of ice floes.

Clearly, flow properties of powders are of great significance for the transfer, sampling, and mixing of pharmaceutical materials. In addition, researchers are interested in using powder flow to probe interactions between particles within a powder. The underlying physical phenomenon of interparticulate interactions have been correlated to Carr's compressibility index (Carr, 1965), Hausner's ratio (Hausner, 1967), and other methods powder flow: Kawakita's constant (Lüdde & Kawakita, 1966) and including shear cell measurement (Carr & Walker, 1967), critical orifice diameter (Walker, 1966) and rotating drums (Castellanos, Valverde, & Quintanilla, 2002; Concessio, VanOort, Knowles, & Hickey, 1999; Crowder & Hickey, 2006; Faqih et al., 2006; Lee, Poynter, Podczeck, & Newton, 2000). Static measures of powder flow give measurable parameters that describe a powder's ability to flow under certain conditions. It is possible to predict the flow behavior of the powder, but the prediction depends on the relationship between each method of quantification and the process being predicted. For this reason, dynamic methods of powder flow have been developed such as flow during vibration, dynamic flow using texture analyzers and strain gauges, and rotating drum methods.

During powder flow the powder must be expanded, and interparticulate forces must be overcome. Particles then increase their separation distance allowing the static system to become more like a fluid. The flow of powders may be characterized in four phases—plastic solid, inertial, fluidization, and suspension (Crowder, Hickey, Louey, & Orr, 2003)—and these phases correlate to the particle spacing, interparticulate forces, and degree of mobility of individual particles in the system. Powders are composed of millions of particles, each of different morphology, size, and other physical properties, and observing flow can be interpreted as random. However, these systems have been suggested to have underlying order during flow (Hickey & Concessio, 1996). Apparent disorder in a chaotic system can be due to a large number of unstable periodic motions (Grebogi, Ott, & Yorke, 1988). In early studies on pharmaceutically relevant powders, Hickey and Concessio showed that using a rotating drum underlying order in powder flow could be detected. As a powder sample is slowly rotated in a drum, the powder rises until its angle of repose is exceeded and an avalanche occurs (Hickey & Concessio, 1996). One can measure the time between avalanching and the dynamic angle of repose. The change of angle of repose with respect to the mean angle plotted against time produces an oscillating data plot. A phase space attractor plot (Concessio & Hickey, 1997) can be generated where data points cluster around a central attractor point. The scatter around this attractor point represents a measure of variability in the avalanching behavior. Lower attractor points indicate better flowability (Kaye, 1993). These oscillating data and phase-space attractor plots can be used as indicators of deterministic chaos, through which patterns can distinguish different powder behaviors (e.g., Fig. 3.2).

Related to the process of flow is the ubiquitous process of mixing and blending in the pharmaceutical sciences. The dynamics of mixing and de-mixing is well known to pharmaceutical scientists and is considered a problem in most cases as it is often hard to predict when a multicomponent mixture will become homogeneous and when it may change into a segregated state. Clearly this has implications for the overarching theme of uniformity of the dosage form that is required of pharmaceutical systems (Shah, Badawy, Szemraj, Gray, & Hussain, 2007; Venables & Wells, 2001). In many cases de-mixing occurs due to size segregation. Particles may reorganize via the gaps found around larger particles, through which smaller particles may slip during energy input into the powder (vibrations, movement, etc.). The lower density of smaller particles at the top of a powder bed then allows the large particles to move upward. In addition, segregation has been well characterized for systems where components have different densities and can be seen in pharmaceutical systems and is now being investigated from a theoretical perspective during manufacturing and processing (Egermann, Krumphuber, & Frank, 1992; Xie et al., 2008). The general approach to ensure uniform mixing or avoidance of segregation in pharmaceutical processing has been through judicious selection of excipients and processing conditions without development of models or assessment of the dynamics of the system. Therefore, recent developments in the mathematics of nonlinear dynamical systems used to describe spontaneous pattern forming, "emergence," and self-assembly have enabled improved understanding and potentially control over mixing. Spontaneous chaotic mixing of powders less than 300 microns diameter was first reported by Shinbrot, Alexander, Moakher, and Muzzio, (1999). Prior to

Fig. 3.2 Lorenz attractor, an example of an attracting set, which has zero measure in the embedding phase space and has fractal dimension and evolves over time (Anonymous, 2010)

Table 3.3 Other applications of pharmaco-complexity in formulation development.

Problem	Reference
Particle morphology and dissolution	Concessio and Hickey (1997), Dokoumetzidis and Macheras (2006), and Farin and Avnir (1992)
Emergence of order in oscillated powders	Moon, Swift, and Swinney (2004)
Fluidized beds	Daw et al. (1995)
Milling	Manai, Delogu, & Rustici (2002)
Controlled drug delivery	Li and Siegel (2000)
Mixing	Christov, Ottino, and Lueptow, (2010), Ottino and Khakhar (2002), and Shinbrot et al. (1999)
Tablet compaction	Leuenberger, Leu, and Bonny (1992)

this report, it was thought that mixing in granular flows was thought to be diffusive. Periodic slipping and sticking of the powders blended in cylindrical tumblers is the mechanism by which this chaotic mixing occurs and allows much more rapid mixing rates to occur (Table 3.3).

References

Ahmad, A. M., Douglas Boudinot, F., Barr, W. H., Reed, R. C., & Garnett, W. R. (2005). The use of Monte Carlo simulations to study the effect of poor compliance on the steady state concentrations of valproic acid following administration of enteric-coated and extended release divalproex sodium formulations. *Biopharmaceutics & Drug Disposition, 26*(9), 417–425.

Akbarieh, M., Dubuc, B., & Tawashi, R. (1987). Surface studies of calcium oxalate dihydrate single crystals during dissolution in the presence of urine. *Scanning Microscopy, 1*(3), 1397–1403.

Akbarieh, M., & Tawashi, R. (1989). Surface studies of calcium oxalate dihydrate single crystals during dissolution in the presence of stone-formers' urine. *Scanning Microscopy, 3*(1), 139–145. discussion 145–136.

Amidon, G. L., Lennernas, H., Shah, V. P., & Crison, J. R. (1995). A theoretical basis for a biopharmaceutic drug classification: The correlation of in vitro drug product dissolution and in vivo bioavailability. *Pharmaceutical Research, 12*(3), 413–420.

Anonymous. Retrieved August 22, 2010., from http://commons.wikimedia.org/wiki/File:Lorenz_system_r28_s10_b2-6666.png.

Aoshima, M., & Satoh, A. (2005). Two-dimensional Monte Carlo simulations of a colloidal dispersion composed of polydisperse ferromagnetic particles in an applied magnetic field. *Journal of Colloid and Interface Science, 288*(2), 475–488.

Avnir, D., Carberry, J. J., Citri, O., Farin, D., Gratzel, M., & AJ, M. E. (1991). Fractal analysis of size effects and surface morphology effects in catalysis and electrocatalysis. *Chaos, 1*(4), 397–410.

Avnir, D., & Farin, D. (1984). Molecular fractal surfaces. *Nature, 308*(5956), 261–263.

Barnsley, M. F., & Rising, H. (1993). *Fractals everywhere*. Boston, MA: Academic Press Professional.

Braatz, R. D., & Hasebe, S. (2002). Particle size and shape control in crystallization processes. *AIChE Symposium, Series: Proceedings of the 6th International Conference on Chemical, Process Control.*

Carr, J. F., & Walker, D. M. (1967). An annular shear cell for granular materials. *Powder Technology, 68*(1), 369–373.

Carr, R. L. (1965). Evaluating flow properties of solids. *Chemical Engineer, 72*, 163–168.

Carstensen, J. T., & Franchini, M. (1993). The use of fractal geometry in pharmaceutical systems. *Drug Development and Industrial Pharmacy, 19*(1–2), 85–100.

Castellanos, A., Valverde, J. M., & Quintanilla, M. A. (2002). Fine cohesive powders in rotating drums: Transition from rigid-plastic flow to gas-fluidized regime. *Physical Review. E, Statistical, Nonlinear, and Soft Matter Physics, 65*(6 Pt 1), 061301.

Christofides, P. D. (2002). *Model-Based Control of Particulate Processes* (Vol. 14). Springer Science & Business Media., Dordrecht, Netherlands.

Christov, I. C., Ottino, J. M., & Lueptow, R. M. (2010). Chaotic mixing via streamline jumping in quasi-two-dimensional tumbled granular flows. *Chaos, 20*(2), 023102.

Concessio, N. M., & Hickey, A. J. (1997). Descriptors of irregular particle morphology and powder properties. *Advanced Drug Delivery Reviews, 26*(1), 29–40.

Concessio, N. M., VanOort, M. M., Knowles, M. R., & Hickey, A. J. (1999). Pharmaceutical dry powder aerosols: Correlation of powder properties with dose delivery and implications for pharmacodynamic effect. *Pharmaceutical Research, 16*(6), 828–834.

Crowder, T., & Hickey, A. (2006). Powder specific active dispersion for generation of pharmaceutical aerosols. *International Journal of Pharmaceutics, 327*(1–2), 65–72.

Crowder, T., Hickey, A., Louey, M. D., & Orr, N. (2003). *A guide to pharmaceutical particulate science*. New York, NY: Informa Healthcare..

D'Arcy, D. M., Corrigan, O. I., & Healy, A. M. (2005). Hydrodynamic simulation (computational fluid dynamics) of asymmetrically positioned tablets in the paddle dissolution apparatus: Impact on dissolution rate and variability. *The Journal of Pharmacy and Pharmacology, 57*(10), 1243–1250.

D'Arcy, D. M., Corrigan, O. I., & Healy, A. M. (2006). Evaluation of hydrodynamics in the basket dissolution apparatus using computational fluid dynamics–dissolution rate implications. *European Journal of Pharmaceutical Sciences, 27*(2–3), 259–267.

Daw, C. S., Finney, C. E. A., Vasudevan, M., van Goor, N. A., Nguyen, K., Bruns, D. D., ... Yorke, J. A. (1995). Self-organization and Chaos in a fluidized bed. *Physical Review Letters, 75*(12), 2308.

Dokoumetzidis, A., Kosmidis, K., Argyrakis, P., & Macheras, P. (2005). Modeling and Monte Carlo simulations in oral drug absorption. *Basic & Clinical Pharmacology & Toxicology, 96*(3), 200–205.

Dokoumetzidis, A., & Macheras, P. (2006). A century of dissolution research: From Noyes and Whitney to the biopharmaceutics classification system. *International Journal of Pharmaceutics, 321*(1–2), 1–11.

Dressman, J. B., Amidon, G. L., Reppas, C., & Shah, V. P. (1998). Dissolution testing as a prognostic tool for oral drug absorption: Immediate release dosage forms. *Pharmaceutical Research, 15*(1), 11–22.

Egermann, H., Krumphuber, A., & Frank, P. (1992). Novel approach to estimate quality of binary random powder mixtures: Samples of constant volume. III: Range of validity of equation. *Journal of Pharmaceutical Sciences, 81*(8), 773–776.

Faqih, A., Chaudhuri, B., Alexander, A. W., Davies, C., Muzzio, F. J., & Silvina Tomassone, M. (2006). An experimental/computational approach for examining unconfined cohesive powder flow. *International Journal of Pharmaceutics, 324*(2), 116–127.

Farin, D., & Avnir, D. (1992). Use of fractal geometry to determine effects of surface morphology on drug dissolution. *Journal of Pharmaceutical Sciences, 81*(1), 54–57.

Fini, A., Fazio, G., Fernández-Hervás, M. J., Holgado, M. A., & Rabasco, A. M. (1996). Fractal analysis of sodium cholate particles. *Journal of Pharmaceutical Sciences, 85*(9), 971–975.

Fini, A., Holgado, M. A., Rodriguez, L., & Cavallari, C. (2002). Ultrasound-compacted indomethacin/polyvinylpyrrolidone systems: Effect of compaction process on particle morphology and dissolution behavior. *Journal of Pharmaceutical Sciences, 91*(8), 1880–1890.

Fuite, J., Marsh, R., & Tuszyński, J. (2002). Fractal pharmacokinetics of the drug mibefradil in the liver. *Physical Review. E, Statistical, Nonlinear, and Soft Matter Physics, 66*(2 Pt 1), 021904.

Fujiwara, M., Nagy, Z. K., Chew, J. W., & Braatz, R. D. (2005). First-principles and direct design approaches for the control of pharmaceutical crystallization. *Journal of Process Control, 15*(5), 493–504.

Gilchrist, J. F., & Ottino, J. M. (2003). Competition between chaos and order: Mixing and segregation in a spherical tumbler. *Physical Review. E, Statistical, Nonlinear, and Soft Matter Physics, 68*(6 Pt 1), 061303.

Grebogi, C., Ott, E., & Yorke, J. A. (1988). Unstable periodic orbits and the dimensions of multifractal chaotic attractors. *Physical Review A, 37*(5), 1711.

Hausner, H. H. (1967). Friction conditions in a mass of metal powder. *International Journal of Powder Metallurgy, 3*, 7–13.

Hickey, A., & Concessio, N. M. (1996). Chaos in rotating lactose beds. *Particulate Science and Technology, 14*(1), 15–25.

Higuchi, T. (1961). Rate of release of medicaments from ointment bases containing drugs in suspension. *Journal of Pharmaceutical Sciences, 50*, 874–875.

Hill, K. M., Khakhar, D. V., Gilchrist, J. F., McCarthy, J. J., & Ottino, J. M. (1999). Segregation-driven organization in chaotic granular flows. *Proceedings of the National Academy of Sciences of the United States of America, 96*(21), 11701–11706.

Holgado, M. A., Fernández-Hervás, M. J., Rabasco, A. M., & Fini, A. (1995). Characterization study of a diclofenac salt by means of SEM and fractal analysis. *International Journal of Pharmaceutics, 120*(2), 157–167.

Jartti, T. T., Kuusela, T. A., Kaila, T. J., Tahvanainen, K. U., & Välimäki, I. A. (1998). The dose-response effects of terbutaline on the variability, approximate entropy and fractal dimension of heart rate and blood pressure. *British Journal of Clinical Pharmacology, 45*(3), 277–285.

Johns, M. L., & Gladden, L. F. (2000). Probing ganglia dissolution and mobilization in a water-saturated porous medium using MRI. *Journal of Colloid and Interface Science, 225*(1), 119–127.

Jorgensen, W. L., & Duffy, E. M. (2000). Prediction of drug solubility from Monte Carlo simulations. *Bioorganic & Medicinal Chemistry Letters, 10*(11), 1155–1158.

Karalis, V., & Macheras, P. (2002). Drug disposition viewed in terms of the fractal volume of distribution. *Pharmaceutical Research, 19*(5), 696–703.

Kaye, B. H. (1978). Specification of the ruggedness and/or texture of a fine particle profile by its fractal dimension. *Powder Technology, 21*(1), 1–16.

Kaye, B. H. (1993). *Chaos & complexity : Discovering the surprising patterns of science and technology*. Weinheim, Germany, New York, NY: VCH.

Kaye, B. H. (1994). *A random walk through fractal dimensions*. Weinheim, Germany/New York, NY: VCH.

Keller, T. H., Pichota, A., & Yin, Z. (2006). A practical view of 'druggability'. *Current Opinion in Chemical Biology, 10*(4), 357–361.

Khakhar, D. V., McCarthy, J. J., Gilchrist, J. F., & Ottino, J. M. (1999). Chaotic mixing of granular materials in two-dimensional tumbling mixers. *Chaos, 9*(1), 195–205.

Kosmidis, K., Argyrakis, P., & Macheras, P. (2003). A reappraisal of drug release laws using Monte Carlo simulations: The prevalence of the Weibull function. *Pharmaceutical Research, 20*(7), 988–995.

Kosmidis, K., & Macheras, P. (2007). Monte Carlo simulations for the study of drug release from matrices with high and low diffusivity areas. *International Journal of Pharmaceutics, 343*(1–2), 166–172.

Kosmidis, K., & Macheras, P. (2008). Monte Carlo simulations of drug release from matrices with periodic layers of high and low diffusivity. *International Journal of Pharmaceutics, 354*(1–2), 111–116.

Kosmidis, K., Rinaki, E., Argyrakis, P., & Macheras, P. (2003). Analysis of Case II drug transport with radial and axial release from cylinders. *International Journal of Pharmaceutics, 254*(2), 183–188.

Kuu, W. Y., & Chilamkurti, R. (2003). Determination of in-process limits during parenteral solution manufacturing using Monte Carlo simulation. *PDA Journal of Pharmaceutical Science and Technology, 57*(4), 263–276.

Lee, Y. S., Poynter, R., Podczeck, F., & Newton, J. M. (2000). Development of a dual approach to assess powder flow from avalanching behavior. *AAPS PharmSciTech, 1*(3), E21.

Leuenberger, H., Leu, R., & Bonny, J. D. (1992). Application of percolation theory and fractal geometry to tablet compaction. *Drug Development and Industrial Pharmacy, 18*(6–7), 723–766.

Li, B., & Siegel, R. A. (2000). Global analysis of a model pulsing drug delivery oscillator based on chemomechanical feedback with hysteresis. *Chaos, 10*(3), 682–690.

Lipinski, C. A., Lombardo, F., Dominy, B. W., & Feeney, P. J. (2001). Experimental and computational approaches to estimate solubility and permeability in drug discovery and development settings. *Advanced Drug Delivery Reviews, 46*(1–3), 3–26.

Liu, J. G., & Nie, Y. F. (2001). Fractal scaling of effective diffusion coefficient of solute in porous media. *Journal of Environmental Sciences (China), 13*(2), 170–172.

Lüdde, K. H., & Kawakita, K. (1966). Die Pulverkompression. *Pharmazie, 21*, 393–403.

Luerkens, D. W. (1991). *Theory and application of morphological analysis : Fine particles and surfaces*. Boca Raton, FL: CRC Press.

Macheras, P. (1996). A fractal approach to heterogeneous drug distribution: Calcium pharmacokinetics. *Pharmaceutical Research, 13*(5), 663–670.

Macheras, P., & Argyrakis, P. (1997). Gastrointestinal drug absorption: Is it time to consider heterogeneity as well as homogeneity? *Pharmaceutical Research, 14*(7), 842–847.

Manai, G., Delogu, F., & Rustici, M. (2002). Onset of chaotic dynamics in a ball mill: Attractors merging and crisis induced intermittency. *Chaos, 12*(3), 601–609.

Marsh, R. E., & Riauka, T. A. (2007). Modeling fractal-like drug elimination kinetics using an interacting random-walk model. *Physical Review. E, Statistical, Nonlinear, and Soft Matter Physics, 75*(3 Pt 1), 031902.

Marsh, R. E., & Tuszynski, J. A. (2006). Fractal Michaelis-Menten kinetics under steady state conditions: Application to mibefradil. *Pharmaceutical Research, 23*(12), 2760–2767.

Martin-Landrove, M., Pereira, D., Caldeira, M. E., Itriago, S., & Juliac, M. (2007). Fractal analysis of tumoral lesions in brain. *Conference Proceedings: Annual International Conference of the IEEE Engineering in Medicine and Biology Society, 2007*, 1306–1309.

Mihranyan, A., & Stromme, M. (2007). Solubility of fractal nanoparticles. *Surface Science, 601*(2), 315–319.

Montgomery, M. J., Beringer, P. M., Aminimanizani, A., Louie, S. G., Shapiro, B. J., Jelliffe, R., & Gill, M. A. (2001). Population pharmacokinetics and use of Monte Carlo simulation to evaluate currently recommended dosing regimens of ciprofloxacin in adult patients with cystic fibrosis. *Antimicrobial Agents and Chemotherapy, 45*(12), 3468–3473.

Moon, S. J., Swift, J. B., & Swinney, H. L. (2004). Role of friction in pattern formation in oscillated granular layers. *Physical Review E, 69*(3), 031301.

Muzzio, F. J., Goodridge, C. L., Alexander, A., Arratia, P., Yang, H., Sudah, O., & Mergen, G. (2003). Sampling and characterization of pharmaceutical powders and granular blends. *International Journal of Pharmaceutics, 250*(1), 51–64.

Narambuena, C. F., Ausar, F. S., Bianco, I. D., Beltramo, D. M., & Leiva, E. P. (2005). Aggregation of casein micelles by interactions with chitosans: A study by Monte Carlo simulations. *Journal of Agricultural and Food Chemistry, 53*(2), 459–463.

Ottino, J. M., & Khakhar, D. V. (2002). Open problems in active chaotic flows: Competition between chaos and order in granular materials. *Chaos, 12*(2), 400–407.

Pang, K. S., Weiss, M., & Macheras, P. (2007). Advanced pharmacokinetic models based on organ clearance, circulatory, and fractal concepts. *The AAPS Journal, 9*(2), E268–E283.

Papadopoulou, V., Kosmidis, K., Vlachou, M., & Macheras, P. (2006). On the use of the Weibull function for the discernment of drug release mechanisms. *International Journal of Pharmaceutics, 309*(1–2), 44–50.

Peppas, N. A. (1985). Analysis of Fickian and non-Fickian drug release from polymers. *Pharmaceutica Acta Helvetiae, 60*(4), 110–111.

Pereira, L. M. (2010). Fractal pharmacokinetics. *Computational and Mathematical Methods in Medicine, 11*(2), 161–184.

Persson, E. M., Gustafsson, A. S., Carlsson, A. S., Nilsson, R. G., Knutson, L., Forsell, P., … Abrahamsson, B. (2005). The effects of food on the dissolution of poorly soluble drugs in human and in model small intestinal fluids. *Pharmaceutical Research, 22*(12), 2141–2151.

Rawlings, J. B., Miller, S. M., & Witkowski, W. R. (1993). Model identification and control of solution crystallization processes: A review. *Industrial & Engineering Chemistry Research, 32*(7), 1275–1296.

Roncaglia, R., Mannella, R., & Grigolini, P. (1994). Fractal properties of ion channels and diffusion. *Mathematical Biosciences, 123*(1), 77–101.

Rowe, R. C., York, P., Colbourn, E. A., & Roskilly, S. J. (2005). The influence of pellet shape, size and distribution on capsule filling–a preliminary evaluation of three-dimensional computer simulation using a Monte-Carlo technique. *International Journal of Pharmaceutics, 300*(1–2), 32–37.

Sanz, E., & Marenduzzo, D. (2010). Dynamic Monte Carlo versus Brownian dynamics: A comparison for self-diffusion and crystallization in colloidal fluids. *The Journal of Chemical Physics, 132*(19), 194102.

Schroder, M., & Kleinebudde, P. (1995). Structure of disintegrating pellets with regard to fractal geometry. *Pharmaceutical Research, 12*(11), 1694–1700.

Shah, K. R., Badawy, S. I., Szemraj, M. M., Gray, D. B., & Hussain, M. A. (2007). Assessment of segregation potential of powder blends. *Pharmaceutical Development and Technology, 12*(5), 457–462.

Shinbrot, T., Alexander, A., Moakher, M., & Muzzio, F. J. (1999). Chaotic granular mixing. *Chaos, 9*(3), 611–620.

Siepmann, J., & Peppas, N. A. (2001). Modeling of drug release from delivery systems based on hydroxypropyl methylcellulose (HPMC). *Advanced Drug Delivery Reviews, 48*(2–3), 139–157.

Siepmann, J., & Siepmann, F. (2013). Mathematical modeling of drug dissolution. *International Journal of Pharmaceutics, 453*(1), 12–24.

Sjoberg, B., & Mortensen, K. (1997). Structure and thermodynamics of nonideal solutions of colloidal particles: Investigation of salt-free solutions of human serum albumin by using small-angle neutron scattering and Monte Carlo simulation. *Biophysical Chemistry, 65*(1), 75–83.

Tromelin, A., Gnanou, J. C., Andrès, C., Pourcelot, Y., & Chaillot, B. (1996). Study of morphology of reactive dissolution interface using fractal geometry. *Journal of Pharmaceutical Sciences, 85*(9), 924–928.

Tromelin, A., Hautbout, G., & Pourcelot, Y. (2001). Application of fractal geometry to dissolution kinetic study of a sweetener excipient. *International Journal of Pharmaceutics, 224*(1–2), 131–140.

Venables, H. J., & Wells, J. I. (2001). Powder mixing. *Drug Development and Industrial Pharmacy, 27*(7), 599–612.

Walker, D. M. (1966). An approximate theory for pressures and arching in hoppers. *Chemical Engineering Science, 21*, 975–997.

Wang, T. Y., Sheng, Y. J., & Tsao, H. K. (2009). Donnan potential of dilute colloidal dispersions: Monte Carlo simulations. *Journal of Colloid and Interface Science, 340*(2), 192–201.

Warnken, Z., Smyth, H. D. C., & Williams, R. O. (2016). Route-specific challenges in the delivery of poorly water-soluble drugs. In R. O. Williams III et al. (Eds.), *Formulating poorly water soluble drugs* (AAPS advances in the pharmaceutical sciences series) (Vol. 22, pp. 1–39).

Weidler, P. G., Degovics, G., & Laggner, P. (1998). Surface roughness created by acidic dissolution of synthetic goethite monitored with SAXS and N2-adsorption isotherms. *Journal of Colloid and Interface Science, 197*(1), 1–8.

Weisstein, E. W. (2010). "Fractal." Retrieved August 18, 2010, from http://mathworld.wolfram.com/Fractal.html.

Xie, L., Wu, H., Shen, M., Augsburger, L. L., Lyon, R. C., Khan, M. A., … Hoag, S. W. (2008). Quality-by-design (QbD): Effects of testing parameters and formulation variables on the segregation tendency of pharmaceutical powder measured by the ASTM D 6940-04 segregation tester. *Journal of Pharmaceutical Sciences, 97*(10), 4485–4497.

Zook, J. M., & Iftekharuddin, K. M. (2005). Statistical analysis of fractal-based brain tumor detection algorithms. *Magnetic Resonance Imaging, 23*(5), 671–678.

Chapter 4
Considerations in Monitoring and Controlling Pharmaceutical Manufacturing

Abstract Monitoring and control of complex processes involve a number of variables whose interactions are necessarily complex. Unlike many other areas of pharmaceutical development, this has long been recognized by process engineers who have the task of guaranteeing the quality and performance of the product. A variety of statistical, physical, and mathematical approaches have been adopted depending on the needs of the assessment. As tools for this purpose have evolved from the ability to manage and store data, the concept of quality by design (QbD) has gained ground and is now a central theme for industry and government regulators. QbD requires significant preparatory consideration of any process by a team of qualified individuals to map out all of the known variables that might contribute to desired attributes of the product. Since there are many variables involved in manufacturing processes and some may not be known, or not subject to control, the mathematical approach to the complexity has included artificial neural networks which have the capacity to learn from data generated and to integrate that knowledge into a predictive approach to a range of activities from research to development. Indeed, advanced mathematical modeling and control tools will be needed as the industry slowly moves from batch to continuous processing.

Keywords Pharmaceutical processing · Continuous processing · Quality by design · Artificial neural networks · Process control · In silico modeling · Design of experiments

Pharmaceutical manufacturing is by definition a complex activity and is perhaps one of the more thoroughly studied activities from this perspective in product development (Hickey & Ganderton, 2010).

A primary objective of the quality by design (QbD) approach that has relatively recently been implemented through guidances and recommendations from pharmaceutical regulatory agencies is to improve process capability and efficiency and reduce process variability through enhanced understanding and control of the process (Brunaugh & Smyth, 2018; Yu, 2008; Yu et al., 2014). This is addressed by identifying the critical quality attributes of the product and ensuring that these attributes are controlled by building quality into the design of the process responsible for them, rather than through testing of the final product. The design of robust

© American Association of Pharmaceutical Scientists 2020

A. J. Hickey, H. D.C. Smyth, *Pharmaco-complexity*, AAPS Introductions in the Pharmaceutical Sciences, https://doi.org/10.1007/978-3-030-42783-2_4

processes that can tolerate known or expected variability in inputs without compromise to the system is thus essential for a quality product. However, modeling, controlling, and understanding many pharmaceutical manufacturing processes can be very challenging due to nonlinearity and significant dependence on initial conditions. To partially address these concerns and limitations of classical pharmaceutical manufacturing, there has recently been a shift from conventional batch manufacturing to continuous processing. In addition to potential benefits of enhanced sustainability, reliability, and cost-effectiveness, continuous manufacturing can enable access to controlled output products which would otherwise be poorly controlled when implemented as batches.

Modeling of processing methods may help predict the processing conditions and allow for simulation of scenarios in which the process causes the pharmaceutical product to become "out of specification." A recent review of process optimization and modeling of milling processes provides an example of the range of mathematical and statistical approaches that can be employed (Brunaugh & Smyth, 2018). In silico modeling based on first principles, experimental data, and combinations of the two are possible. In the milling of pharmaceutical solids, for example, modeling based on energy laws, population-based method (PBM) approaches, discrete element method (DEM) approaches, and computational fluid dynamic (CFD) simulations have all been reported.

4.1   Optimizing Processes Using Statistics and Experimental Design

Cochran and Cox (1957) were among the first to describe statistical approaches to experimental design. By implementing sound statistical principles, products with desired attributes can be prepared efficiently and reproducibly.

4.1.1   Sources of Error

The sources of experimental error are varied but can be considered and ultimately controlled with appropriate experimental design and subsequent data analysis. Randomization provides a basis for experimental design that overcomes coincidental effects and allows causation to be inferred. Effects are frequently complex and do not conform to linearity or simple additive interpretation. Selected experimental designs allow for interactive and nonlinear effects to be identified without confounding with experimental error.

4.1.2 Randomized and Latin Square Designs

Assigning treatments to units within the design randomly is the simplest way to map data for analysis. This approach affords flexibility as any number of replicates and treatments may be employed; statistical ease of analysis even if replicates for some units or whole treatments are lost. Loss of information from missing data is less than other designs. However, this approach may lead to loss of accuracy resulting from the fact that the whole variation is uniformly distributed across treatments and units and becomes part of the experimental error. Different designs may be employed to reduce the error. For example, randomized blocks may be useful. Initially the design calls for dividing the experimental product into groups, which represent a single trial or replication. For example, product could be blocked for sampling time to assign specific error to environmental conditions. Latin square designs group treatments in replicates. The design assigns treatments to rows or columns. This eliminates errors from differences among rows and columns. Consequently, the Latin square design reduces error compared with random blocks.

4.1.3 Factorial Design

Multiple variables can be evaluated concurrently by adopting a factorial design approach. The simplest approach evaluates each factor at two levels (2^n factorial design). For example, a simple jet milling process requires consideration of two opposing gas pressures and time. Studying each factor at a low and high level results in a 2^3 factorial design.

Factorial experiments frequently investigate the effects of each factor over some pre-designated range encompassed by the levels of that factor and are not intended to identify the combination of factors yielding the minimum or maximum response. Where factors are independent, statistical analysis is easy. Additional information may be gained through confounding analysis if it is believed that factors are not independent. Factorial analysis is a rapid and efficient method for identifying an operating process space that is likely to allow product quality and performance attributes to be achieved.

4.1.4 Fractional Factorial Design

Full factorial designs for complex processes may be too costly and time-consuming. In addition, the precision obtained may be far beyond that required for decision-making. In a 2^7 factorial design, each main effect is an average of 64 combinations of other factors. It may be sufficient to conduct an 8- or 16-fold replication. Information in this partial approach is lost in particular with respect to interactions between factors, and so some caution should be exercised. A screening study may benefit from this approach as it has little impact, but a foundational study on which serious decisions would be based might require greater attention.

4.1.5   Central Composite Design

The previous designs considered linear functions between factors. Where nonlinear functions are considered, the simplest approach is a quadratic response surface (second order) obtained from a central composite designs based on factorial analysis. Central composite designs (CCD) test additional factors and their combinations and can be fitted sequentially program of experimentation. Starting with an exploratory 2^n factorial design where the center of the first experiment is close to a point of maximum response, combinations of factors may be picked orthogonally to indicate the curvature of the response surface.

4.1.6   Response Surface Maps

CCDs can be extended to a broad range of combinations of factors and levels to obtain a continuous nonlinear surface that predicts the response to factor variation. For the CCD example, the response surface assessment began hypothetically when the process was near the optimum. Beyond this point it would be desirable to approach the true response function at a small defined region around the optimum. It is important to acknowledge that the true response is curved at this location. Sequential experiments should be performed within a region of variable space, known as the operability region (OR). These experiments allow mapping over a particular region of interest, response optimization, and selection of operating conditions required to meet specifications.

4.2   Process Design and Control

4.2.1   Quality by Design

A rational strategy of process or response evaluation is essential to an understanding and to the creation of knowledge. The efficiency and reproducibility of engineering processes are a challenge to the robustness of experimental deigns and their ability to probe and optimize operating conditions due to the overall level of complexity.

Quality by design encompasses a rage of techniques to proactively address all parameters scientifically to mitigate the risk of not meeting product quality and performance requirements and considers the design activity from conception to market.

Fig. 4.1 Ishikawa
(fishbone) approach to
capturing input factors and
relationship to output
parameters. (Modified
from Hickey & Ganderton,
2010)

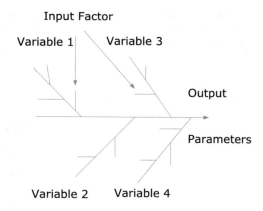

Tools are available to guide the product development team deliberations and include a variety of branched factor relationship diagram including the fishbone (Ishikawa) diagram. This diagram is intended to facilitate the analytical process for the group.

A fishbone diagram is illustrated in Fig. 4.1. Variables are considered in terms of the way in which they impinge on a process as assessed by a predetermined output parameter. Regardless of the stage of assessment and the purpose of the exercise, a thorough preliminary understanding of the process under consideration should result from this approach.

A thorough assessment of the process before entering experimentation increases the potential of identifying all important variables; instills confidence that all factors have been considered, through the breadth of expertise minimizes potential to overlook factors; reduces time and cost of experimental missteps; and ultimately allows rational design of process space. Appropriate statistical methods as described earlier are required ideally with the ability for real-time monitoring and control through process analytical technology.

4.2.2   Design Space

Single response surface map and actual process design space may be distinguished by the latter being dynamic beginning when drug is conceived and evolving through the product life cycle (Lepore & Spavins, 2008).

The mathematical principles of the statistical approach to experimental design would be hard to do justice in brief, but fortunately there are excellent texts on this topic (Box, Hunter, & Hunter, 1978). Figure 4.2 provides an example of process design space development which is dynamic and will evolve over the time course of the drug product development.

Fig. 4.2 Design space development. (Modified from Hickey & Ganderton, 2010)

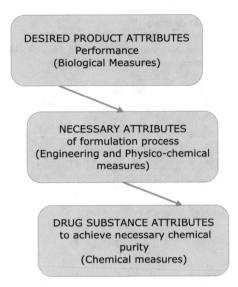

Process Analytical Technology

To exert control over any process in pharmaceutical product development, its response to changes in manufacturing variables must be monitored. Where batch production methods are employed, batch sampling followed by assessment of particular properties (drug content, particle size, etc.) must be conducted. The production conditions are then controlled to achieve designated quality specifications.

Quantitative analytical methods have improved to now allow real-time, in-process measurements to be taken, and feedback control systems allow adjustments in input parameters to achieve continuous monitoring and control of processes. A detailed discussion of process analytical technology (PAT) can be found elsewhere (Hickey & Ganderton, 2010). A range of processes have been studied with respect to applying principles of quality by design all in the arena of engineering such as crystallization, drying, and milling.

4.3   Risk Assessment and Management

The concept of criticality is central to risk assessment and management. There are three designations with regard to criticality: noncritical, critical, and an undesignated low-risk category. Variables that have not been shown to impact on safety or efficacy or factor into critical quality attributes (CQAs) as defined by ICH Q(8) R are not considered critical. Consequently, they do not have to be included in design space. Critical variables are those that are known to impact safety, efficacy, or other measures of biological disposition or compliance. Critical process parameters if varied outside a particular range directly and significantly influence on CQAs. These properties must be controlled within predefined range and are thought to ensure final product quality. The intermediate undesignated category with regard to criticality represents attributes that may impact the product but represent a low risk. The term "low risk" is based on an indirect impact on safety and/or efficacy alone or in combination with other variables; mitigated risk; knowledge transfer from noncritical variables requires additional evaluation.

Criticality may be reduced to the elements of severity, occurrence, and detection in a compounding manner (Nosal & Schultz, 2008) which in turn can be related to experimental design (frequency and variation) and analytical capability (detection). Clear differentiation of levels of criticality is required during the life cycle of the product to facilitate a control strategy for process variables, material attributes, and their contribution to quality measures.

Given the complex nature of pharmaceutical processes and manufacturing in general, there have been attempts to adopt methods in mathematical complexity to assist in more efficiently and reproducibly guaranteeing the quality and performance of these systems. Given the likelihood that all variables are not known to the investigator, the approach that has been most common does not invoke a model that definitively fits the conditions but rather assumes that it can be educated by knowledge of experimental results. The application of artificial neural networks (ANNs) has been addressed in detail elsewhere in the context of pharmaceutical research (Agatanovic-Kustrin & Berseford, 2000), sciences (Achanta, Kowalski, & Rhodes, 1995), formulation (Takayama, Fujikawa, & Nagai, 1999), and development (Borquin, 1997). It suffices to say that to this point these approaches have not superseded those of quality by design supported by statistical method, but it is to be hoped that more work on this subject will occur in the coming years.

4.4 Summary

The complexity of pharmaceutical manufacturing is a challenge to control. However, unlike many other aspects of pharmaceutical product development, discussed in other sections, the problem has been so severe as to warrant significant research leading to practical approaches that support industry and regulatory standards. The basic principles behind these controls are statistically designed experiments to support definition of design space with superimposed designations for critical parameters that impinge on the quality and performance of the drug. Forays into the application of mathematical complexity in this field have begun but are limited at this point to research exercises that will hopefully extend to practical application in the future.

References

Agatanovic-Kustrin, S., & Berseford, R. (2000). Basic concepts of artificial neural network (ANN) modeling and its application to pharmaceutical research. *Journal of Pharmaceutical and Biomedical Analysis, 22,* 717–727.

Achanta, A. S., Kowalski, J. G., & Rhodes, C. T. (1995). Artificial neural networks: Implications for pharmaceutical sciences. *Drug Development and Industrial Pharmacy, 21*, 119–155.

Borquin, J. (1997). Basic concepts of artificial neural networks (ANN) modeling in the application to pharmaceutical development. *Pharmaceutical Development and Technology, 2*, 95–109.

Box, G. E. P., Hunter, W. G., & Hunter, J. S. (1978). *Statistics for experimenters: An introduction to design, data analysis, and model building*. New York, NY: John Wiley and Sons.

Brunaugh, A., & Smyth, H. D. C. (2018). Process optimization and particle engineering of micronized drug powders via milling. *Drug Delivery and Translational Research, 8*(6), 1740–1750.

Cochran, W. G., & Cox, G. M. (1957). *Experimental designs* (2nd ed.). New York, NY: John Wiley and Sons.

Hickey, A. J., & Ganderton, D. (2010). *Pharmaceutical process engineering* (2nd ed.). New York, NY: Informa Healthcare.

Lepore, J., & Spavins, J. (2008). PQLI design space. *Journal of Pharmaceutical Innovation, 3*, 79–87.

Nosal, R., & Schultz, T. (2008). PQLI definition of criticality. *Journal of Pharmaceutical Innovation, 3*, 69–78.

Takayama, K., Fujikawa, M., & Nagai, T. (1999). Artificial neural network (ANN) as a novel method to optimize pharmaceutical formulation. *Pharmaceutical Research, 16*, 1–6.

Yu, L. X. (2008). Pharmaceutical quality by design: Product and process development, understanding, and control. *Pharmaceutical Research, 25*(4), 781–791.

Yu, L. X., Amidon, G., Khan, M. A., Hoag, S. W., Polli, J., Raju, G. K., et al. (2014). Understanding pharmaceutical quality by design. *The AAPS Journal, 16*, 771–783.

Chapter 5
Fractal Pharmacokinetics, Systems Pharmacology, Network Analysis, Multiscale, Kinetics, Toxicokinetics, Drug Safety

Abstract The complexity of pharmacokinetics and drug efficacy during clinical development is due to the multiscale understanding required from molecular drug-target interactions to organismal-level phenotypes. Since the first edition of this volume, a steady growth in advanced mathematical and/or computational techniques has been reported in the improved description and understanding of pharmacokinetics and pharmacodynamics. In addition to fractal pharmacokinetics, systems pharmacology and network analysis have provided researchers with better tools to describe drug kinetics in the treatment of disease. These and other approaches are summarized in this chapter.

Keywords Fractal pharmacokinetics · Systems pharmacology · Network analysis · Multi scale · Kinetics · Toxicokinetics · Drug safety

Many physiological systems appear to present data that is random or without order. The origins of this disorder are often attributed to variability introduced by the multifactorial determinants of the system. This classical view of physiological randomness has also been widely discussed in quantitative pharmacological systems. However, since the early 1990s, it has been revealed that highly variable data from physiological, pharmacokinetic, and pharmacodynamic studies, in contrast to errors in measurement, can have their origins in nonlinear dynamical systems that can be described by chaos theory (Goldberger, 1989, 1996; Goldberger, Rigney, & West, 1990; Tallarida, 1990a, 1990b; van Rossum & de Bie, 1991; Dokoumetzidis, Iliadis, & Macheras, 2001, 2002; Mager & Abernethy, 2007).

5.1 Pharmacokinetics/Toxicokinetics

The interaction of drugs and exogenous substances with the body, and, specifically, the time course of these compounds within the body, falls in the general field of pharmacokinetics (i.e., "what the body does to the drug"). Absorption, distribution,

A. J. Hickey, H. D.C. Smyth, *Pharmaco-complexity*, AAPS Introductions in the Pharmaceutical Sciences, https://doi.org/10.1007/978-3-030-42783-2_5

metabolism, and excretion are all processes that occur sequentially and simultaneously upon administration of a drug/toxicant to the organism. The pharmacokineticist must assess these processes via sampling of body fluids to determine drug concentrations and the subsequent use of pharmacokinetic models. There is often a relationship between plasma drug concentrations and pharmacologic responses, but these are also often complex. Interpretation of pharmacokinetic data involves mathematical modeling to enable prediction of blood/plasma and tissue concentrations time profiles and perhaps facilitate mechanistic understanding of the underlying processes. It is widely recognized that these models currently used are vast oversimplifications, "but we have to start somewhere" (Pang, Weiss, & Macheras, 2007).

Pharmacokinetics must predict the safety and efficacy of drugs and is a major component of the development and regulation of pharmaceuticals. Drug developers must use pharmacokinetics to determine the optimal dose where the most benefit is obtained with the least risk of potential side effects. Around 95% of new drugs have been reported to have suboptimal pharmacokinetic profiles (Orive, Gascon, Hernández, Domínguez-Gil, & Pedraz, 2004). Drug discovery via high-throughput screening has also tended to produce candidate drugs with physicochemical properties that compromise pharmacokinetic profiles (e.g., poor or erratic absorption). Although pharmacokinetics is seen as a critical for its predictive power, the tools widely employed to enable optimal prediction may not represent reality closely enough. Moreover, it might be imagined that given the suboptimal or complex pharmacokinetic profiles, delivery systems may be designed, via a more comprehensive understanding of the underlying pharmacokinetic mechanisms, to maximize the drug benefit while minimizing risk.

Classical PK is reduced to linear relationships but actually is modeled by exponential (log) functions. It is a very similar phenomenon to chemical kinetics in that it can be modeled by linearity, but the underlying complexity which is more informative is lost in the averaging phenomenon (i.e., Chap. 2).

5.1.1 Compartmental Modeling

Of the various approaches used to predict drug time profiles in the body, compartmental models are the simplest and most widely used (Pang et al., 2007). The basis of the compartment model is equilibrium systems, homogeneity, and mass transfer between compartments. A compartment is defined as the number of drug molecules having the same probability of undergoing a set of chemical kinetic processes (Marsh & Tuszynski, 2006). The mass transfer of drug molecules between compartments is described by kinetic rate constants. As alluded to above, these models assume that each compartment is homogenous (instantaneous mixing) and mass transfer constants are constant. Therefore, the system can be described by coupled first-order differential equations. Macheras and colleagues argue that the assumptions of homogeneity and well-stirred media are not supported by anatomical and physiological evidence (Macheras, Argyrakis, & Polymilis, 1996; Macheras & Argyrakis, 1997; Dokoumetzidis, Karalis, Iliadis, & Macheras, 2004; Pang et al., 2007).

5.1.2   Fractal Kinetics

The realization that physiological and anatomical characteristics of the human body are complex and not well described by compartment models has lead several researchers to investigate pharmacokinetics from a fractal point of view. The concept of fractals was introduced in earlier sections. The argument made is that diffusion, a major component of drug transport and kinetics, is not well described by Fick's laws in the human body because of under-stirred areas, and constrained spaces are known to exist. Furthermore, biological systems are made up of a multitude of interacting parts and will be nonlinear from a dynamical systems point of view (Dokoumetzidis et al., 2004). As mentioned previously in this text, fractals have the property of being self-similar at various scales of scrutiny. In physiology, fractals can be found in numerous organ systems, for example, the vascular tree, the lungs, and the folds of the brain, which have all been well described using fractal geometry. Application of fractals to pharmacokinetics has been used for drug release and dissolution (as shown in Chap. 3), absorption (Macheras & Argyrakis, 1997), and fractal compartment models (Fuite, Marsh, & Tuszyński, 2002; Marsh & Tuszynski, 2006). Excellent reviews of these fractal concepts applied to pharmacokinetics have been published, and the field continues to expand (Dokoumetzidis et al., 2004; Pang et al., 2007; Sopasakis, Sarimveis, Macheras, & Dokoumetzidis, 2018).

5.1.3   Physiologically Relevant Modeling

Physiologically based pharmacokinetic (PBPK) models relate organ or tissue structures to the physiology of the organ or tissue. The PBPK models can be considered as a data-informed approach to achieve a more biologically realistic dose-response model. PBPK models facilitate the estimation of drug concentrations at target and off-target tissues by taking into account the rate of absorption into the body, distribution and storage in tissues, metabolism, and excretion on the basis of interplay among critical physiological, physicochemical, and biochemical determinants. Qualitative evidence that this type of modeling is more relevant can be observed from those agencies and institutions involved in risk assessment of exposure. For example, physiologically based pharmacokinetic models have increasingly been employed in chemical health risk assessments carried out by the US Environmental Protection Agency (EPA), and it is anticipated that their use will continue to increase. Relevant physiological parameter values (e.g., alveolar ventilation, blood flow and tissue volumes, glomerular filtration rate) are critical components of these models, and values can now be found in various databases for use by researchers and risk assessors. From a mathematical point of view, a PBPK model comprises a system of coupled ordinary differential equations (ODEs). These equations involve physiological and physicochemical parameters, each of which is typically affected with uncertainty and some degree of variability due to inter- and intraindividual variations. To investigate the effects of initial values and uncertainty in parameters for these models (and the ODEs), Monte Carlo methods have been used (e.g., Thomas, Bigelow, Keefe, and Yang (1996)).

5.1.4   Systems Pharmacology and Network Analysis

More recently, a systems pharmacology approach has been combined with PBPK models (e.g., Danhof, 2016). In a systems approach, a network of connected nodes are used to help describe functional interactions. This may allow better explanations of fundamental properties of biological systems behavior such as hysteresis, nonlinearity, variability, interdependency, convergence, resilience, and multi-stationarity (Danhof, 2016). The systems pharmacology approach is hoped to improve drug development. For example, the clinical efficacy failure of molecules that have shown therapeutic promise in cellular and animal models arises from the lack of understanding of human biology/pathophysiology which incorporates multiscale mechanisms spanning from molecular-level drug-target interactions to organismal-level phenotypes (Zhao & Iyengar, 2012).

5.2   Efficacy/Safety

In linear systems, the result of an input, such as the efficacy or toxicity of a drug, is proportional to the stimulus (e.g., dose), and multiple stimuli result in the summation of the inputs. These systems are quite attractive to those wishing to understand and predict the effects of stimuli (i.e., responses to a certain dose) as classical mathematics and models that can describe this type of linearity are very familiar to us. However, in nonlinear systems, the resulting behavior of a series of inputs is not equal to the summation of all the components and individual behaviors. These basic principles of complex systems and chaos have now been widely reported and popularized. However, application of these methods of interpretation and modeling of pharmacokinetic and pharmacodynamic data are not commonplace. This, no doubt, is due to the unfamiliarity of the mathematical methods and the absence of standard training in this field for biologists and pharmaceutical scientists. It is clear, however, physiological systems are nonlinear. The advantage of nonlinear systems for physiological processes is suggested to be one of energy efficiency and control compared to linear systems (Dokoumetzidis et al.,

2001). To predict the future state of a dynamical system requires iterating the mathematical function many times, advancing time in small steps. This iteration procedure is referred to as "solving the system" or integrating. Using fast computer processing now readily available, it is possible to determine all future points using just the initial point. This time series is known as the trajectory or orbit. In nonlinear dynamical systems, the data appears to be random and seemingly unpredictable and has been described as chaos. The mathematics related to nonlinear dynamical systems and chaos focuses not on "solving the system" and finding precise solutions to the governing equations (as this often appears hopeless) but rather on determining qualitatively how the system will behave in the long term. This is because chaotic systems are very sensitive to small differences in initial conditions; i.e., errors due to very small rounding errors in numerical computation may result in widely diverging outcomes. Thus, despite the fact that chaotic systems are deterministic (their future of the system is fully determined by the initial conditions, not random behavior), the high sensitivity of the system on initial conditions makes long-term prediction of outcomes very difficult and often impossible. In summary, chaotic systems are deterministic but not predictable. A classical example is the weather (Smith, 2002), but chaos can also be found within trivial systems, and therefore it is not surprising that physiological responses to drug administration may also be described by nonlinear dynamics and chaos.

Classical pharmacodynamic models and approaches are unable to adequately deal with such complex and variable data. Ligand-receptor interactions, upon which traditional pharmacodynamics is largely based, have been shown to be complex (Prank et al., 1995). The classical Emax model widely used in pharmacodynamic studies (Schoemaker, van Gerven, & Cohen, 1998) may not sufficiently deal with deviations that are commonly observed in practice, such as feedback mechanisms induced by ligand-receptor interactions that nonlinearly influence the pharmacodynamic response (Dokoumetzidis et al., 2001). In fact, many examples are readily available from closely related fields:

(a) Physiological responses to hormones and neurotransmitters: These endogenous substances interact with cellular receptors in controlled reactions governed by the mass action law, and some subsets of receptors result in negative feedback that reduces further release rates of the ligand (Tallarida & Freeman, 1994). The system was shown to pass from periodic to chaotic as the input parameters were varied.

(b) Cytokines, small protein or glycoprotein messenger molecules, allow transfer of information from one cell to another. There are large numbers of different types of cytokines that can trigger complex intracellular signaling cascades in other cells. Individual cytokines have multiple and diverse biological functions. A panoply of feedback loops operate within these complex cytokine-cellular systems that result in complex nonlinear behavior (Callard, George, & Stark, 1999; Seely & Christou, 2000; Higgins, 2002).

Additional examples are found in an excellent review by Dokoumetzidis et al. (2001).

Mager and Abernethy (2007) cite that the biological variability in biological signaling and the complexity of pharmacological systems often complicate or preclude the direct application of traditional structural and nonstructural models in pharmacodynamics. They suggest that mathematical transforms (such as fast Fourier and wavelet transforms) of data sets may be used to provide better measures of drug effects and patterns in responses and to assist in interpretation of diverse data sets (e.g., imaging and biomedical signals). As observed in our earlier example of powder flow analysis, mathematical transforms may also be required for pharmacodynamic data interpretation as the time series data is nonlinear and the frequency content of a signal is more informative than the original waveform.

5.3   Summary

Methods dealing with complexity in PK/PD studies are key to establishing how drugs should be administered, which is a risk analysis task, achieved by defining and analyzing the potential benefits and dangers to individuals and populations. In practice, new chemical entities are evaluated in vitro, in preclinical models in vivo, then in human using clinical trials. Early trials are focused on safety and establishing potential risks of the drug. The larger-scale clinical trials focus on efficacy but also must gather data on safety. These studies are analyzed statistically, and approval is generally based on envisioning the average patient. However, it is clear that current methods may not be able to predict or guide drug development at these later phases to the extent desired.

References

Callard, R., George, A. J., & Stark, J. (1999). Cytokines, chaos, and complexity. *Immunity, 11*(5), 507–513.

Danhof, M. (2016). Systems pharmacology – Towards the modeling of network interactions. *European Journal of Pharmaceutical Sciences, 94*, 4–14.

Dokoumetzidis, A., Iliadis, A., & Macheras, P. (2001). Nonlinear dynamics and chaos theory: Concepts and applications relevant to pharmacodynamics. *Pharmaceutical Research, 18*(4), 415–426.

Dokoumetzidis, A., Iliadis, A., & Macheras, P. (2002). Nonlinear dynamics in clinical pharmacology: The paradigm of cortisol secretion and suppression. *British Journal of Clinical Pharmacology, 54*(1), 21–29.

Dokoumetzidis, A., Karalis, V., Iliadis, A., & Macheras, P. (2004). The heterogeneous course of drug transit through the body. *Trends in Pharmacological Sciences, 25*(3), 140–146.

Fuite, J., Marsh, R., & Tuszyński, J. (2002). Fractal pharmacokinetics of the drug mibefradil in the liver. *Physical Review. E, Statistical, Nonlinear, and Soft Matter Physics, 66*(2 Pt 1), 021904.

Goldberger, A. L. (1989). Cardiac chaos. *Science, 243*(4897), 1419.

Goldberger, A. L. (1996). Non-linear dynamics for clinicians: Chaos theory, fractals, and complexity at the bedside. *Lancet, 347*(9011), 1312–1314.

Goldberger, A. L., Rigney, D. R., & West, B. J. (1990). Chaos and fractals in human physiology. *Scientific American, 262*(2), 42–49.

Higgins, J. P. (2002). Nonlinear systems in medicine. *The Yale Journal of Biology and Medicine, 75*(5–6), 247–260.

Macheras, P., & Argyrakis, P. (1997). Gastrointestinal drug absorption: Is it time to consider heterogeneity as well as homogeneity? *Pharmaceutical Research, 14*(7), 842–847.

Macheras, P., Argyrakis, P., & Polymilis, C. (1996). Fractal geometry, fractal kinetics and chaos en route to biopharmaceutical sciences. *European Journal of Drug Metabolism and Pharmacokinetics, 21*(2), 77–86.

Mager, D. E., & Abernethy, D. R. (2007). Use of wavelet and fast Fourier transforms in pharmacodynamics. *The Journal of Pharmacology and Experimental Therapeutics, 321*(2), 423–430.

Marsh, R. E., & Tuszynski, J. A. (2006). Fractal michaelis-menten kinetics under steady state conditions: Application to mibefradil. *Pharmaceutical Research, 23*(12), 2760–2767.

Orive, G., Gascon, A. R., Hernández, R. M., Domínguez-Gil, A., & Pedraz, J. L. (2004). Techniques: New approaches to the delivery of biopharmaceuticals. *Trends in Pharmacological Sciences, 25*(7), 382–387.

Pang, K. S., Weiss, M., & Macheras, P. (2007). Advanced pharmacokinetic models based on organ clearance, circulatory, and fractal concepts. *The AAPS Journal, 9*(2), E268–E283.

Prank, K., Harms, H., Brabant, G., Hesch, R. D., Dammig, M., & Mitschke, F. (1995). Nonlinear dynamics in pulsatile secretion of parathyroid hormone in normal human subjects. *Chaos, 5*(1), 76–81.

Schoemaker, R. C., van Gerven, J. M., & Cohen, A. F. (1998). Estimating potency for the Emax-model without attaining maximal effects. *Journal of Pharmacokinetics and Biopharmaceutics, 26*(5), 581–593.

Seely, A. J., & Christou, N. V. (2000). Multiple organ dysfunction syndrome: Exploring the paradigm of complex nonlinear systems. *Critical Care Medicine, 28*(7), 2193–2200.

Smith, L. A. (2002). What might we learn from climate forecasts? *Proceedings of the National Academy of Sciences of the United States of America, 99*(Suppl 1), 2487–2492.

Sopasakis, P., Sarimveis, H., Macheras, P., & Dokoumetzidis, A. (2018). Fractional calculus in pharmacokinetics. *Journal of Pharmacokinetics and Pharmacodynamics, 45*, 107–125.

Tallarida, R. J. (1990a). Control and oscillation in ligand receptor interactions according to the law of mass action. *Life Sciences, 46*(22), 1559–1568.

Tallarida, R. J. (1990b). Further characterization of a control model for ligand-receptor interaction: Phase plane geometry, stability, and oscillation. *Annals of Biomedical Engineering, 18*(6), 671–684.

Tallarida, R. J., & Freeman, K. A. (1994). Chaos and control in mass-action binding of endogenous compounds. *Annals of Biomedical Engineering, 22*(2), 153–161.

Thomas, R. S., Bigelow, P. L., Keefe, T. J., & Yang, R. S. (1996). Variability in biological exposure indices using physiologically based pharmacokinetic modeling and Monte Carlo simulation. *American Industrial Hygiene Association Journal, 57*(1), 23–32.

van Rossum, J. M., & de Bie, J. E. (1991). Chaos and illusion. *Trends in Pharmacological Sciences, 12*(10), 379–383.

Zhao, S., & Iyengar, R. (2012). Systems pharmacology: Network analysis to identify multiscale mechanisms of drug action. *Annual Review of Pharmacology and Toxicology, 52*(1), 505–521.

Chapter 6
Impact of Complexity on Population Biology

Abstract The complexity of biological systems is recognized superficially, but there has been a tendency through reductionism to believe that fundamental understanding is achieved through examination of the smallest building blocks of life. There is steadily increasing understanding that looking at large populations particularly as the tools have become available to probe the underpinning rules of genetics and epigenetics will lead to a systematic understanding that may offer unique strategies for future disease therapy. Since the first edition of this book, many of the predictions with respect to unraveling the biological complexity through genomics, transcriptomics, metabolomics, and proteomics have come to pass, and a host of new therapies particularly for rare diseases are under development.

Keywords Population biology · Disease · Genetic disorders · Epidemiology · Adverse events

The advent of whole genome sequencing has opened a new era in therapeutic agent discovery and development (Consortium, 2001). Transcriptomics, proteomics, and metabolomics have made significant advances in the two decades since the genome was deciphered (Horgan & Kenny, 2011). As a consequence, there is a large body of knowledge on the function of specific proteins and other biological molecules that are important to life (Boycott, Vanstone, Bulman, & Mac Kenzie, 2013). In addition, the high incidence of rare diseases frequently associated with monogenic disorders may illuminate whole organism functionality (Berman, 2014). A systems biology approach to reconstruct the high-level functionality through interactive biochemical and biophysical networks is rapidly becoming the foundation of modern medicine (Westerhoff et al., 2009).

The previous chapters have demonstrated that the physical chemistry, engineering, and consequently regulatory considerations required to developed pharmaceutical products when scrutinized closely involve complex and confounding variables that require a high level of control.

Clinically it is recognized that human organisms, and their homeostatic mechanisms, are also very complex, but historically little has been done to recognize this in approaches to therapy. As novel discoveries in the biological sciences and biotechnology are brought to bear on disease, revolutionary new treatments are likely to be developed.

© American Association of Pharmaceutical Scientists 2020
A. J. Hickey, H. D.C. Smyth, *Pharmaco-complexity*, AAPS Introductions in the Pharmaceutical Sciences, https://doi.org/10.1007/978-3-030-42783-2_6

6.1   Nature of Disease

Populations of people are diverse with respect to phenotype in terms of age, gender, ethnicity, and also geographical location (important for environmental stimuli/ exposure, diet, etc.) and also genotype which underpins the phenotype and results in different responses to aberrant cell/tissue types or pathogens. The additional layer of complexity related to the proteome and metabolomics or regulation of gene expression and metabolism renders the biological system very complex. Consequently, the predictive nature of treatment or prevention of disease is subject to enormous variability.

At the beginning of the millennium, the first publications on whole human genome sequences opened the possibility of revolutionary approaches to disease treatment and prevention (Levy, Sutton, Ng, Feuk, & Halpern, 2007; Venter, Adams, Myers, Li, & Muraj, 2001). Initially the pharmaceutical industry believed that mapping single nucleotide polymorphisms would indicate important new targets for drug therapy (Altshuler, Pollara, Cowles, Etten, & Baldwin, 2000). However, as rare diseases become a focus for new therapeutic approaches, compilations of individual genomes of affected individuals seem to be more likely to result in identification of targets (Turnbull, Ahmed, Morrison, Pernet, & Renwick, 2010). The complex biology that results in translation and expression of genes into the major functional building block proteins makes the proteome an important and diverse component of the interaction of an organism with its environment (O'Donovan, Apweiler, & Baroch, 2001). Indeed, since functionality is key to understanding, prediction, and control of disease, a large emphasis has been placed on a sub-category of proteins that are involved in metabolomics (Pearson, 2007).

6.1.1   Genetic Disorders

Since the whole human genome was characterized, the incidence of genetic disorders, which may previously have been noted as having unknown etiology, has emphasized the nature of control and integration of function that supports life and health (Dawkins et al., 2018). An alarmingly small number of therapeutic agents are available to treat these disorders. New therapies are being developed to correct mutations with gene therapy, to correct function by targeting with small molecular weight molecules, or to ameliorate the biochemical deficiency with enzyme replacement therapy.

As more is learned about the spectrum of rare diseases, it may be possible to group them according to function. For example, there are many diseases that are characterized by mitochondrial dysfunction or lysosomal storage (Garone & Viscomi, 2018; Glew, Basu, Prence, & Remaley, 1985; Winchester, Vellodi, & Young, 2000). Understanding rare diseases in the context of common diseases may be the key to unlocking systems functionality behind health (Quitaina-Murci, 2016; Sarntivijal et al., 2016).

Clearly, the amount of data that will be accumulated to address the link between molecular scale mutations and whole organism functionality will be enormous. The complexity of this problem will only be manageable through computer-assisted strategies (Davis et al., 2016; Kanehisa et al., 2014; Krumholz, 2014).

6.2   Potential for Disease Intervention

There has been a considerable interest in mathematical approaches to the population biology of infectious diseases for many years (RH Anderson & Francis, 2018; RM Anderson & May, 1979). The ability to rapidly sequence infectious microorganisms, such as *Mycobacterium tuberculosis* (Cole, Brosch, Parkhill, Garnier, & Churcher, 1998), put researchers in a unique position to consider bioinformatic approaches to map genome, proteome, and metabolome of infectious microorganisms as therapeutic targets.

In the first instance, the outcome of these new discoveries and the evolution of a longstanding interest in population biology are targeted therapeutic strategies for specific diseases, such as cancer (Gatenby & Vincent, 2003). However, the logical outcome of the developments in molecular and cellular biology is the possibility of individualized therapies and preventative strategies (Evans & Relling, 2004).

The demand for more unique approaches to therapy will feedback to the properties of drugs, dosage forms, and their manufacture requiring much greater flexibility and speed of product development (Hirano, 2007). This will only be achieved by maximizing the data collection (as is already occurring in molecular biology) and subjecting it to rigorous scrutiny to yield as much interpretable information and knowledge as possible to facilitate decision-making and efficient pharmaceutical project and risk management (Kennedy, 1998; Vesper, 2006).

6.2.1   Epidemiological Studies and Adverse Events

Epidemiological studies investigating causal hypotheses often are inconsistent from one study to the next dependent on methodology. Contradictory results may be explained through complexity and nonlinearity (Glattre & Nygard, 2004). The possibly fractal nature of ordered series of relative risks (RR) and their possible self-organized criticality (SOC) have been suggested. Using these concepts, Glattre and Nygard performed reanalysis of published meta-studies, one of which investigates the possible association of oral contraceptives and female breast cancer and found different conclusions than those made by the original study.

The incidence of adverse drug reactions is typically reported in a statistical method, but few studies have, from a population standpoint, investigated the variability of responses or the root cause of why they occur. Frattarelli showed that, with limitations, the severity of adverse drug reactions follows a distribution seen in other complex adaptive systems, called a power law distribution, and that preventable reactions occurred for reasons fundamentally different from those that underlie the nonpreventable reactions (Dokoumoetzidis & Macheras, 2006; Frattarelli, 2005). Using published studies, the author plotted incidence of drug reaction as a function of severity and then performed a fit to an equation. Overall and nonpreventable drug reactions followed a power law distribution regardless of sample size or the nature of the population or drugs studied. Preventable reactions, on the other hand, were described by a different type of equation.

6.3 Summary

The enormous strides that have been made in understanding human and pathogen genomes, proteomes, and metabolomes give the medical and pharmaceutical community a unique opportunity in the history of mankind's pursuit of knowledge to truly develop targeted therapies to address populations and subpopulations of people for variants on particular diseases. Moreover, taken to its logical conclusion, the ability to differentiate the particular manifestations of disease associated with individual therapies tailored to their specific needs becomes a distinct possibility. Having followed the theme of this small volume, it should be evident that such a development would put enormous pressure on pharmaceutical scientists and engineers to adapt to a rapidly changing product development and manufacturing environment to bring all of the relevant monitoring and controls of the complex phenomena described earlier to bear to support this radical new approach to disease therapy and prevention.

References

Altshuler, D., Pollara, V., Cowles, C., Etten, W. V., & Baldwin, J. (2000). An SNP map of the human genome generated by reduced representation shotgun sequencing. *Nature, 407*, 513–516.

Anderson, R., & Francis, K. (2018). Modeling rare diseases with induced pluripotent stem cell technology. *Molecular and Cellular Probes, 40*, 52–59.

Anderson, R., & May, R. (1979). Population biology of infectious disease:Part I. *Nature, 280*, 361–367.

Berman, J. (2014). *Rare diseases and orphan drug: Keys to understanding*. Waltham, MA: Academic Press.

Boycott, K., Vanstone, M., Bulman, D., & MacKenzie, A. E. (2013). Rare-disease genetics in the era of next-generation sequencing: Discovery to translation. *Nature Reviews Genetics, 14*, 681–691.

Cole, S., Brosch, R., Parkhill, R., Garnier, T., & Churcher, C. (1998). Deciphering the biology of Mycobacterium tuberculosis from the complete genome sequence. *Nature, 393*, 537–544.

Davis, A., Wiegers, T., King, B., Wiegers, J., Grondin, C., Sciaky, D., … Mattingly, C. (2016). Generating gene ontology-disease inferences to explore mechanisms of human disease at the comparative toxicogenomics database. *PLoS One, 11*(5), e0155530.

Dawkins, H., Draghia-Akli, R., Lasko, P., Lau, L., Jonker, A., Cutillo, C., ... Austin, C. (2018). Progress in rare diseases research 2010-2016: An IRDiRC perspective. *Clinical and Translational Science, 11*(1), 11–20.

Dokoumoetzidis, A., & Macheras, P. (2006). A comment on "adverse drug reactions and avalanches: Life on the edge of chaos". *Journal of Clinical Pharmacology, 46*(9), 1057–1058. (author reply 1059-1060).

Evans, W., & Relling, M. (2004). Moving toward individualizeed medicine with pharmacogenomics. *Nature, 429*, 464–468.

Frattarelli, D. (2005). Adverse drug reactions and avalanches: Life at the edge of chaos. *Journal of Clinical Pharmacology, 45*(8), 866–871.

Garone, C., & Viscomi, C. (2018). Towards a therapy for mitochondrial disease: An update. *Biochemical Society Transactions, 46*(5), 1247–1261.

Gatenby, R., & Vincent, T. (2003). Application of quantitative models from population biology and evolutionary game theory to tumor therapeutic strategies. *Molecular Cancer Therapeutics, 2*, 919–927.

Glattre, E., & Nygard, J. (2004). Fractal meta-analysis and causality embedded in complexity: Advanced understanding of disease etiology. *Nonlinear Dynamics, Psychology, and Life Sciences, 8*(3), 315–344.

Glew, R., Basu, A., Prence, E., & Remaley, A. (1985). Lysosomal storage diseases. *Laboratory Investigation, 53*(3), 250–269.

Hirano, T. (2007). Cellular pharmacodynamics of immunosuppressive drugs for indivudualized medicine. *International Immunopharamcol, 7*, 3–22.

Horgan, R., & Kenny, L. (2011). 'Omic' technologies: Genomics, transcriptomics, proteomics and metabolomics. *The Obstetrician & Gynaecologist, 13*(3), 189.

Human Genome Sequencing Consortium. (2001). Initial sequencing and analysis of the human genome. *Nature, 409*, 860–921.

Kanehisa, M., Goto, S., Sato, Y., Kawashima, M., Furumichi, M., & Tanabe, M. (2014). Data, information, knowledge and principle: Back to metabolism in KEGG. *Nucleic Acid Research, 42*(D1), D199–D205.

Kennedy, T. (1998). *Pharmaceutifcal project management*. New York, NY: Marcel Dekker, Inc..

Krumholz, H. (2014). Big data and new knowledge in medicine: The thinking, training, and tools needed for a learning health system. *Health Affairs (Millwood), 33*(7), 1163–1170.

Levy, S., Sutton, G., Ng, P., Feuk, L., & Halpern, A. (2007). The diploid genome sequence of an individual human. *PLoS Biology, 5*(10), e254.

O'Donovan, C., Apweiler, R., & Baroch, A. (2001). The human proteomics initiative. *Trends in Biotechnology, 19*, 178–181.

Pearson, H. (2007). Meet the human metabolome. *Nature, 446*, 8. https://doi.org/10.1038/446008a

Quitaina-Murci, L. (2016). Understanding rare and common diseases in the context of human evolution. *Genome Biology, 17*, 225.

Sarntivijal, S., Vasant, D., Jupp, S., Saunders, G., Bento, A., Gonzalez, D., ... Malone, J. (2016). Linking rare and common disease: Mapping clinical disease-phenotypes to ontologies in therapeutic target validation. *Journal of Biomedical Semantics, 7*, 8.

Turnbull, C., Ahmed, S., Morrison, J., Pernet, D., & Renwick, A. (2010). Genome-wide association study identifies five new breast cancer susceptible loci. *Nature Genetics, 42*, 504.

Venter, J., Adams, M., Myers, E., Li, P., & Muraj, R. (2001). The sequence of the human genome. *Sicnece, 291*, 1304–1351.

Vesper, J. (2006). *Risk assessment and risk management in the pharmaceutical industry: Clear and simple*. Washington, DC: Parenteral Drug Association.

Westerhoff, H., Winder, C., Messiha, H., Simeonidis, E., Adamczyk, M., Verma, M., ... Dunn, W. (2009). Systems biology: The elements and principles of life. *FEBS Letters, 584*(24), 3882–3890.

Winchester, B., Vellodi, A., & Young, E. (2000). The molecular basis for lysosomal storage diseases and their treatment. *Biochemical Society Transactions, 28*(2), 150–154.

Chapter 7
Computational Modeling of Nonlinear Phenomena Using Machine Learning

Abstract Machine learning (ML) is a field of computer science that allows interrogation to allow modified navigation (learning) of the data and through statistical derivation prediction of unseen data or events. ML has been a high-profile topic for many years and is ubiquitous in many aspects of daily life – from e-mail spam and malware filtering to search results refining online customer service and fraud detection. More recently, ML has been pervasive in solving complex nonlinear phenomena in pharmaceutical and medical sciences. It has been used in modeling chemical data sets for two decades. It has only recently become a useful approach to improve healthcare diagnoses and to provide personalized medical treatments. The rapid growth in data collection and integration, as well as the accessibility of increasing computing power, especially in cloud services, explains this unforeseen capacity to transform data into information, information into knowledge, and knowledge into wisdom (see Fig. 7.1). In this section, we briefly introduce the concepts and types of ML and its application for drug discovery, drug product development, and clinical application. The literature in these fields and the importance and challenges of interpreting ML results are also discussed.

Keywords Computational modeling · Machine learning · Artificial intelligence · Drug discovery · Product development · Clinical application

7.1 Concepts and Applications

As noted previously in this volume, most of the pharmacological and biological processes are nonlinear. Linear models deal with correlation. Indirect relationships happening through multiple co-occurrences can be modeled through second linear models. However, as discussed in this book, most biological and pharmacological processes are nonlinear in nature. This means that these processes cannot be modeled by simple correlation, i.e., linear models. Machine learning (ML) and artificial intelligence (AI) have become a practical solution to overcome the human limitation in modeling complex phenomena. In this section, we will briefly introduce these concepts and describe how we can benefit from it to advance to therapeutic lead identification, drug product development, and clinical application as well as to help to interpret these models to derive knowledge from large data sets.

© American Association of Pharmaceutical Scientists 2020
A. J. Hickey, H. D. C. Smyth, *Pharmaco-complexity*, AAPS Introductions in the Pharmaceutical Sciences, https://doi.org/10.1007/978-3-030-42783-2_7

In the last couple of decades, due to the ability to process large amounts of data, ML has transformed the industry, by allowing the development of virtual personal assistants (Kepuska & Bohouta, 2018), such as Siri, Alexa, and Google, predicting commuter traffic (Horvitz, Apacible, Sarin, & Liao, 2012), and even to construct self-driving cars (Bojarski et al., 2016). It has also been employed to predict human behavior based on social media data (Ruths & Pfeffer, 2014) and financial statement fraud detection (Perols, 2011) and to build a machine for automatic detection of retinopathy in diabetic patients using image processing (Decencière et al., 2013), among others.

AI is not new. It was first described in 1955 by John McCarthy (McCarthy, Minsky, Rochester, & Shannon, 1955). Many authors, such as Hubert Dreyfus (1979), discussed the philosophy behind how computers learn and questioned if they would ever be smarter than humans. However, the underlying explanation for the limited application of ML over many decades was the lack of data, which is still a major problem in several fields, such as drug discovery for rare diseases (see the next section). The democratization of computer access, as well as the advances in computer processing power and data storage, has allowed the development of scaled AI technologies applied to large and distributed data sets, which ultimately led to the resurrection and overspread of AI in the last few years [ref].

ML is a branch of AI, and it is a method of data analysis used to build systems that can learn from data. Mitchell has defined ML as "A computer program is said to learn from experience E with respect to some class of tasks T and performance measure P, if its performance at tasks in T, as measured by P, improves with experience E" (Mitchell, 1997). In the context of pharmaco-complexity and drug discovery, we can think of it as the binding affinity to a particular protein, while the experience is the number of compounds tested. As the quantity and quality of data (chemicals) tested increase (Fourches, Muratov, & Tropsha, 2010), so does the ability of the algorithm to increase its performance to predict the binding affinity task (Vamathevan et al., 2019). ML is used to automate model building, which can be used to identify patterns and make predictions that assist humans in making decisions.

Usually, ML algorithms can be divided into supervised, unsupervised, semi-supervised, and reinforcement. The supervised algorithm works with labeled data, i.e., the dependent variable is known, and the algorithms will work to learn from the independent variables (predictors) to better describe the labels. This task can be either classificatory or regression. On the other hand, unsupervised algorithms, such as clustering methods, study the data to identify patterns. Semi-supervised use labeled and unlabeled data to predict the unlabeled data. Lastly, reinforcement learning algorithms are based on trial-and-error procedures, which learn from past experience, evaluate them, and adapt the algorithm to improve its performance (Goodfellow, Bengio, & Courville, 2016).

The algorithms can also be divided into deterministic and stochastic (probabilistic). In the first, the parameters that modify the predictors of the model are exact, as well as the output. In stochastic models, there is some level of randomness within the parameters, which may vary the output as well (Renard, Alcolea, &

Ginsbourger, 2013). Random forests (Breiman, 2001) are an example of a stochastic algorithm. However, this level of randomness can be controlled by setting a random seed. Recently, deep neural networks, or deep learning (DL), has revolutionized the industry due to the ability of the algorithm to learn from stochastic events by approximating arbitrary nonrandom input-output mappings (Huval et al., 2015). For instance, the use of stochastic gradient descent iterating through a randomly selected subset of the data allows the optimization of the model for large-scale problems (Bottou, 2010).

7.2 Employing Machine Learning and Artificial Intelligence for Drug Discovery, Product Development, and Clinical Application

In drug discovery, ML is often employed to predict chemical structures lacking experimental values in a technique named quantitative structure-activity relationship (QSAR). QSAR is a major computational approach widely employed in industry and academia. Its applications include, but are not limited to, the identification and optimization of the bioactive profile of chemicals and the identification of putative toxicants. Here, ML generates a classification or regression rule from a training set of molecules with known biological data (Cherkasov et al., 2014; Dearden, 2016). The algorithm is trained using molecular descriptors, which correspond to numbers generated from a mathematical procedure to describe the chemical structure of the compounds (Todeschini & Consonni, 2009). Molecular descriptors are calculated for chemical compounds with biological data for a determined endpoint. These descriptors are used to train the algorithm. Once validated, the generated models constitute the starting point for the selection of compounds in virtual screening campaigns. The methods of validation for these models have been widely discussed in the literature (Dearden, Cronin, & Kaiser, 2009; Tropsha, 2010).

In the last couple of decades, ML has become a major driver in drug discovery (Ekins et al., 2019; Lavecchia, 2015). In 2013, the now-former President-Elect of the Royal Society of Chemistry forecasted that "in 15 years' time, no chemist will be doing any experiments at the bench without trying to model them first" (Tildesley & Care, 2014). As we write this book, the result of his prediction is yet to be validated. Although difficult to believe that every chemist will be using computational models to drive their research, as we can see in Fig. 7.1, there has been steep growth in the co-occurrence of ML, QSAR, and drug discovery in the literature over the years.

The discovery of drugs through the application of ML is hard to estimate since the drug discovery pipeline is rather a spiral than linear. However, as corroborated by Fig. 7.1, as of November 2019, a search of the terms "QSAR," "discovery," and "chemical" returns 1112 manuscripts in PubMed. If we replace "QSAR" with "machine learning," it returns 588 publications. By looking at a few, we can note that this approach has been shown to identify novel bioactive compounds for many

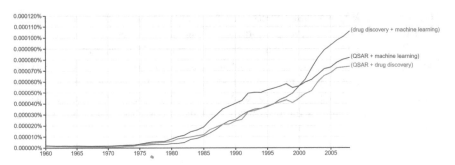

Fig. 7.1 ML has become a major approach used in drug discovery. (The chart was generated by Google Ngram Viewer (http://books.google.com/ngrams)); Y-axis, percentage among all books in the Google Ngram database; X-axis, years

common diseases, such as cancer (Dhiman & Agarwal, 2016; Speck-Planche, 2019; Zhang et al., 2007), type 2 diabetes (Abuhammad & Taha, 2016; Pantaleao et al., 2017), and hypertension (Durdagi, Erol, Dogan, & Berkay Sen, 2019), and also several neglected tropical diseases such as schistosomiasis (Melo Calixto, Braz dos Santos, Clecildo Barreto Bezerra, & de Almeida SilvaID, 2018; Melo-Filho et al., 2016; Neves et al., 2016), malaria (Lima et al., 2018; Neves et al., 2019; Zhang et al., 2013), and Ebola (Capuzzi et al., 2018).

The rapid chemical data growth in public repositories such as ChEMBL (Gaulton et al., 2012) and PubChem (Wang et al., 2012) have also allowed the development of computational models for chemical toxicity assessment. These models have the premise to advance knowledge to derive novel adverse outcome pathway (AOP), which can illustrate relevant toxicity mechanisms (Ciallella & Zhu, 2019). Some toxicity endpoints have been extensively covered in the past few years, such as liver toxicity (Liu et al., 2015; Low et al., 2011; Zhu & Kruhlak, 2014), Ames mutagenicity (Alves, Golbraikh, et al., 2018; Sushko et al., 2010; Xu et al., 2012), and skin sensitization (Alves, Capuzzi, et al., 2018; Dearden et al., 2015; Hisaki, Aiba, Yamaguchi, & Sasa, 2015). And many of the developed models are available for the use of the scientific community through web and stand-alone applications (Alves, Braga, Muratov, & Andrade, 2018; Braga et al., 2017; Xu et al., 2012). However, although these models have been able to predict the toxicity outcome of chemicals correctly, they will only be accepted for regulatory purposes when associated with other testing strategies (Casati et al., 2018; Kleinstreuer et al., 2018).

QSAR technologies have been implemented to advance drug product development. For instance, in a recent study (Alves et al., 2019), the authors developed an innovative computational approach to rationally design polymeric micelle-based formulations to improve the water solubility of poorly soluble drugs. Their approach was validated by an experimental laboratory. Many drugs fail due to poor aqueous solubility, and, therefore, their method can be used to speed up the development of drug candidates and bring new drugs to the market faster. In another interesting study, authors (Fernandez et al., 2019) have estimated developed quantitative structure-price relationship (QS$R) models to assess the economic feasibility of compounds selected in virtual screening campaigns to follow-up into preclinical studies.

Another major objective is to use data-driven approaches with ML to repurpose approved drugs (Zhao & So, 2019). Drug repurposing consists of identifying diseases outside the original therapeutic indication for approved drugs (Ashburn & Thor, 2004). The major benefit of this approach is to skip preclinical and clinical safety evaluation, which could bring the drug much faster and costing less to treat a disease (Nosengo, 2016). This task is performed by using biological data, such as genomics and proteomics data to understand the patterns of diseases and match them with targets and off-targets, i.e., expected and unexpected effects of known drugs (Zhao & So, 2019). A prime example of success of this approach is the story of Matt Might (www.matt.might.net), director of the Hugh Kaul Precision Medicine Institute at the University of Alabama. Matt has a child with a rare disease involving two mutations in NGLY1, and that also led to N-acetylglucosamine, an important amino sugar (Chen, Shen, & Liu, 2010). By employing data analysis and computational methods, he eventually discovered that NGLY1 deficiency could potentially be treated with endo-beta-N-acetylglucosaminidase (ENGase) inhibitors (Bi, Might, Vankayalapati, & Kuberan, 2017). In this context, ML can also be used for personalized medical treatment, i.e., precision medicine, where one could use genetic data from the patient and be used along with chemical and biological data of approved drugs within ML models as well as knowledge graphs for repurposing approved drugs or other supplements for one individual (Ping, Watson, Han, & Bui, 2017).

ML has also become a significant approach to health informatics. During the last few years, DL has been transformative to healthcare, offering the ability to analyze data with unprecedented speed and precision. This is due to the nature of the algorithm, having multiple levels of representation that are obtained by composing simple, but nonlinear, modules (Miotto, Wang, Wang, Jiang, & Dudley, 2017). At each step, the raw input is transformed, and due to the composition of enough transformations, very complex functions can be learned, and outcomes predicted outcomes within vast quantities of unstructured data (LeCun, Bengio, & Hinton, 2015). Healthcare data is very complex since the collected data does not come from and standardized experiment, but unique individuals experiencing a variety of clinical features. Recently, DL has been used to predict 30-day readmission after hospitalization for chronic obstructive pulmonary disease (Goto et al., 2019) and predict patient outcome for a malignant type of cancer (Courtiol et al., 2019) and significant mental health disorders (Graham et al., 2019).

7.3   Interpreting Machine Learning Models to Expand Knowledge of Pharmaco-complexity

An essential benefit of using ML algorithms is to derive insight from uncorrelated variables used to build the model. Although many computational models are often referred to as a "black box" approach (Castelvecchi, 2016), many groups have shown that models could be interpreted (Doshi-Velez & Kim, 2017; Koh & Liang, 2017). Understanding the model is necessary not only to derive knowledge from underlying relationships but also ease the transition from manual to automated processes.

In the context of drug discovery, by interpreting developed models (Polishchuk, Kuz'min, Artemenko, & Muratov, 2013), one can better understand relevant structural moieties interact with the binding site of the molecular target which will ultimately guide the design of novel structures that will modulate the binding affinity of ligands. A practical way to do that is by developing iterative models where the f most important fragments are selected in each iteration, in a method named Chemistry-Wide Association Studies (CWAS) (Low et al., 2018). CWAS was inspired by the GWAS (genome-wide association study), a well-established approach for simultaneously identifying and studying large numbers of genetic features potentially associated with a given phenotype (e.g., disease) (Klein, 2005). In CWAS, the difference is that one would use chemical compounds, molecular descriptors, and biological activity, instead of patients, single-nucleotide polymorphisms, and phenotypes.

The interpretability of DL models has always been a subject of debate (Chakraborty et al., 2017). Although these algorithms tend to achieve higher accuracy, they are more challenging to interpret due to the multilevel layers and the weights transforming the variables (Lipton, 2016). Even though most DL algorithms are deterministic, stochastic ones, such as random forest, are easier to interpret. For this reason, DL has not become popular in fields such as in finance, where many countries require institutions to sue models that are easy to be interpreted by humans, although some approaches have been proposed (Luo, Wu, & Wu, 2017).

Similarly, interpretability is crucial for the broad adoption of computational models in medical research and clinical decision-making (Kerr, Bansal, & Pepe, 2012). Recently, a study used gradient boosting trees to learn interpretable models, eventually achieving the same accuracies than DL on a pediatric intensive care unit data set for acute lung injury (Che, Purushotham, Khemani, & Liu, 2016). Often, there is a need to balance interpretability and accuracy, but this level will depend on the topic of research.

References

Abuhammad, A., & Taha, M. O. (2016). QSAR studies in the discovery of novel type-II diabetic therapies. *Expert Opinion on Drug Discovery, 11*(2), 197–214. https://doi.org/10.1517/17460 441.2016.1118046

Alves, V., Braga, R., Muratov, E., & Andrade, C. (2018). Development of web and mobile applications for chemical toxicity prediction. *Journal of the Brazilian Chemical Society, 29*(5), 982–988. https://doi.org/10.21577/0103-5053.20180013

Alves, V. M., Capuzzi, S. J., Braga, R. C., Borba, J. V. B., Silva, A. C., Luechtefeld, T., ... Tropsha, A. (2018). A perspective and a new integrated computational strategy for skin sensitization assessment. *ACS Sustainable Chemistry & Engineering, 6*(3), 2845–2859. https://doi.org/10.1021/acssuschemeng.7b04220

Alves, V. M., Golbraikh, A., Capuzzi, S. J., Liu, K., Lam, W. I., Korn, D. R., ... Tropsha, A. (2018). Multi-Descriptor Read Across (MuDRA): A simple and transparent approach for developing accurate quantitative structure–activity relationship models. *Journal of Chemical Information and Modeling, 58*(6), 1214–1223. https://doi.org/10.1021/acs.jcim.8b00124

Alves, V. M., Hwang, D., Muratov, E., Sokolsky-Papkov, M., Varlamova, E., Vinod, N., … Kabanov, A. (2019). Cheminformatics-driven discovery of polymeric micelle formulations for poorly soluble drugs. *Science Advances, 5*(6), eaav9784. https://doi.org/10.1126/sciadv. aav9784

Ashburn, T. T., & Thor, K. B. (2004). Drug repositioning: Identifying and developing new uses for existing drugs. *Nature Reviews Drug Discovery, 3*(8), 673–683. https://doi.org/10.1038/ nrd1468

Bi, Y., Might, M., Vankayalapati, H., & Kuberan, B. (2017). Repurposing of Proton Pump Inhibitors as first identified small molecule inhibitors of endo-β-N-acetylglucosaminidase (ENGase) for the treatment of NGLY1 deficiency, a rare genetic disease. *Bioorganic & Medicinal Chemistry Letters, 27*(13), 2962–2966. https://doi.org/10.1016/j.bmcl.2017.05.010

Bojarski, M., Del Testa, D., Dworakowski, D., Firner, B., Flepp, B., Goyal, P., … Zieba, K. (2016). End to end learning for self-driving cars. *ArXiv, 1604.07316.* Retrieved from http://arxiv.org/ abs/1604.07316

Bottou, L. (2010). Large-scale machine learning with stochastic gradient descent. In *Proceedings of COMPSTAT'2010* (pp. 177–186). https://doi.org/10.1007/978-3-7908-2604-3_16

Braga, R. C., Alves, V. M., Muratov, E. N., Strickland, J., Kleinstreuer, N., Tropsha, A., & Andrade, C. H. (2017). Pred-skin: A fast and reliable web application to assess skin sensitization effect of chemicals. *Journal of Chemical Information and Modeling, 57*(5), 1013–1017. https://doi. org/10.1021/acs.jcim.7b00194

Breiman, L. (2001). Random forests. *Machine Learning, 45*(1), 5–32. https://doi.org/10.102 3/A:1010933404324

Capuzzi, S. J., Sun, W., Muratov, E. N., Martínez-Romero, C., He, S., Zhu, W., … Tropsha, A. (2018). Computer-aided discovery and characterization of novel Ebola virus inhibitors. *Journal of Medicinal Chemistry, 61*(8), 3582–3594. https://doi.org/10.1021/acs.jmedchem.8b00035

Casati, S., Aschberger, K., Barroso, J., Casey, W., Delgado, I., Kim, T. S., … Zuang, V. (2018). Standardisation of defined approaches for skin sensitisation testing to support regulatory use and international adoption: Position of the International Cooperation on Alternative Test Methods. *Archives of Toxicology, 92*(2), 611–617. https://doi.org/10.1007/s00204-017-2097-4

Castelvecchi, D. (2016). Can we open the black box of AI? *Nature, 538*(7623), 20–23. https://doi. org/10.1038/538020a

Chakraborty, S., Tomsett, R., Raghavendra, R., Harborne, D., Alzantot, M., Cerutti, F., … Gurram, P. (2017). Interpretability of deep learning models: A survey of results. In *2017 IEEE SmartWorld, Ubiquitous Intelligence & Computing, Advanced & Trusted Computed, Scalable Computing & Communications, Cloud & Big Data Computing, Internet of People and Smart City Innovation (SmartWorld/SCALCOM/UIC/ATC/CBDCom/IOP/SCI)* (pp. 1–6). https://doi. org/10.1109/UIC-ATC.2017.8397411

Che, Z., Purushotham, S., Khemani, R., & Liu, Y. (2016). Interpretable deep models for ICU outcome prediction. In *AMIA … annual symposium proceedings. AMIA symposium, 2016* (pp. 371–380). Retrieved from http://www.ncbi.nlm.nih.gov/pubmed/28269832

Chen, J.-K., Shen, C.-R., & Liu, C.-L. (2010). N-acetylglucosamine: Production and applications. *Marine Drugs, 8*(9), 2493–2516. https://doi.org/10.3390/md8092493

Cherkasov, A., Muratov, E. N., Fourches, D., Varnek, A., Baskin, I. I., Cronin, M., … Tropsha, A. (2014). QSAR modeling: Where have you been? Where are you going to? *Journal of Medicinal Chemistry, 57*(12), 4977–5010. https://doi.org/10.1021/jm4004285

Ciallella, H. L., & Zhu, H. (2019). Advancing computational toxicology in the big data era by artificial intelligence: Data-driven and mechanism-driven modeling for chemical toxicity. *Chemical Research in Toxicology, 32*(4), 536–547. https://doi.org/10.1021/acs.chemrestox.8b00393

Courtiol, P., Maussion, C., Moarii, M., Pronier, E., Pilcer, S., Sefta, M., … Clozel, T. (2019). Deep learning-based classification of mesothelioma improves prediction of patient outcome. *Nature Medicine, 25*(10), 1519–1525. https://doi.org/10.1038/s41591-019-0583-3

Dearden, J. C. (2016). The history and development of quantitative structure-activity relationships (QSARs). *International Journal of Quantitative Structure-Property Relationships, 1*(1), 1–44. https://doi.org/10.4018/IJQSPR.2016010101

Dearden, J. C., Cronin, M. T. D., & Kaiser, K. L. E. (2009). How not to develop a quantitative structure-activity or structure-property relationship (QSAR/QSPR). *SAR and QSAR in Environmental Research, 20*(3–4), 241–266. https://doi.org/10.1080/10629360902949567

Dearden, J. C., Hewitt, M., Roberts, D. W., Enoch, S. J., Rowe, P. H., Przybylak, K. R., … Katritzky, A. R. (2015). Mechanism-based QSAR modeling of skin sensitization. *Chemical Research in Toxicology, 28*(10), 1975–1986. https://doi.org/10.1021/acs.chemrestox.5b00197

Decencière, E., Cazuguel, G., Zhang, X., Thibault, G., Klein, J. C., Meyer, F., … Chabouis, A. (2013). TeleOphta: Machine learning and image processing methods for teleophthalmology. *IRBM, 34*(2), 196–203. https://doi.org/10.1016/j.irbm.2013.01.010

Dhiman, K., & Agarwal, S. M. (2016). NPred: QSAR classification model for identifying plant based naturally occurring anti-cancerous inhibitors. *RSC Advances, 6*(55), 49395–49400. https://doi.org/10.1039/c6ra02772e

Doshi-Velez, F., & Kim, B. (2017). Towards a rigorous science of interpretable machine learning. *ArXiv, 1702.08608.* Retrieved from http://arxiv.org/abs/1702.08608

Dreyfus, H. (1979). *What computers can't do: The limits of artificial intelligence.* London, UK: MIT Press.

Durdagi, S., Erol, I., Dogan, B., & Berkay Sen, T. (2019). Integration of text mining and binary QSAR models for novel anti-hypertensive antagonist scaffolds. *Biophysical Journal, 116*(3), 478a. https://doi.org/10.1016/j.bpj.2018.11.2583

Ekins, S., Puhl, A. C., Zorn, K. M., Lane, T. R., Russo, D. P., Klein, J. J., … Clark, A. M. (2019). Exploiting machine learning for end-to-end drug discovery and development. *Nature Materials, 18*(5), 435–441. https://doi.org/10.1038/s41563-019-0338-z

Fernandez, M., Ban, F., Woo, G., Isaev, O., Perez, C., Fokin, V., … Cherkasov, A. (2019). Quantitative structure–price relationship (QS$R) Modeling and the development of economically feasible drug discovery projects. *Journal of Chemical Information and Modeling, 59*(4), 1306–1313. https://doi.org/10.1021/acs.jcim.8b00747

Fourches, D., Muratov, E., & Tropsha, A. (2010). Trust, but verify: On the importance of chemical structure curation in cheminformatics and QSAR modeling research. *Journal of Chemical Information and Modeling, 50*(7), 1189–1204. https://doi.org/10.1021/ci100176x

Gaulton, A., Bellis, L. J., Bento, A. P., Chambers, J., Davies, M., Hersey, A., … Overington, J. P. (2012). ChEMBL: A large-scale bioactivity database for drug discovery. *Nucleic Acids Research, 40*(Database issue), D1100–D1107. https://doi.org/10.1093/nar/gkr777

Goodfellow, I., Bengio, Y., & Courville, A. (2016). *Deep learning.* Retrieved from http://www.deeplearningbook.org/

Goto, T., Jo, T., Matsui, H., Fushimi, K., Hayashi, H., & Yasunaga, H. (2019). Machine learning-based prediction models for 30-day readmission after hospitalization for chronic obstructive pulmonary disease. *COPD: Journal of Chronic Obstructive Pulmonary Disease*, 1–6. https://doi.org/10.1080/15412555.2019.1688278

Graham, S., Depp, C., Lee, E. E., Nebeker, C., Tu, X., Kim, H.-C., & Jeste, D. V. (2019). Artificial intelligence for mental health and mental illnesses: An overview. *Current Psychiatry Reports, 21*(11), 116. https://doi.org/10.1007/s11920-019-1094-0

Hisaki, T., Aiba, M., Yamaguchi, M., & Sasa, H. (2015). Development of QSAR models using artificial neural network analysis for risk assessment of repeated-dose, reproductive , and developmental toxicities of cosmetic ingredients. *The Journal of Toxicological Sciences, 40*(2), 163–180. https://doi.org/10.2131/jts.40.163

Horvitz, E. J., Apacible, J., Sarin, R., & Liao, L. (2012). Prediction, expectation, and surprise: Methods, designs, and study of a deployed traffic forecasting service. *ArXiv, 1207.1352.* Retrieved from http://arxiv.org/abs/1207.1352

Huval, B., Wang, T., Tandon, S., Kiske, J., Song, W., Pazhayampallil, J., … Ng, A. Y. (2015). An empirical evaluation of deep learning on highway driving. *ArXiv, 1504.01716.* Retrieved from http://arxiv.org/abs/1504.01716

Kepuska, V., & Bohouta, G. (2018). Next-generation of virtual personal assistants (Microsoft Cortana, Apple Siri, Amazon Alexa and Google Home). In *2018 IEEE 8th annual computing and communication workshop and conference, CCWC 2018, 2018-January* (pp. 99–103). https://doi.org/10.1109/CCWC.2018.8301638

Kerr, K. F., Bansal, A., & Pepe, M. S. (2012). Further insight into the incremental value of new markers: The interpretation of performance measures and the importance of clinical context. *American Journal of Epidemiology, 176*, 482–487. https://doi.org/10.1093/aje/kws210

Klein, R. J. (2005). Complement factor H polymorphism in age-related macular degeneration. *Science (New York, N.Y.), 308*(5720), 385–389. https://doi.org/10.1126/science.1109557

Kleinstreuer, N. C., Karmaus, A. L., Mansouri, K., Allen, D. G., Fitzpatrick, J. M., & Patlewicz, G. (2018). Predictive models for acute oral systemic toxicity: A workshop to bridge the gap from research to regulation. *Computational Toxicology, 8*(4), 21–24. https://doi.org/10.1016/j.comtox.2018.08.002

Koh, P. W., & Liang, P. (2017). Understanding black-box predictions via influence functions. In *ICML'17 proceedings of the 34th international conference on machine learning* (pp. 1885–1894). Retrieved from https://dl.acm.org/citation.cfm?id=3305576

Lavecchia, A. (2015). Machine-learning approaches in drug discovery: Methods and applications. *Drug Discovery Today, 20*(3), 318–331. https://doi.org/10.1016/j.drudis.2014.10.012

LeCun, Y., Bengio, Y., & Hinton, G. (2015). Deep learning. *Nature, 521*(7553), 436–444. https://doi.org/10.1038/nature14539

Lima, M. N. N., Melo-Filho, C. C., Cassiano, G. C., Neves, B. J., Alves, V. M., Braga, R. C., ... Andrade, C. H. (2018). QSAR-driven design and discovery of novel compounds with antiplasmodial and transmission blocking activities. *Frontiers in Pharmacology, 9*, 146. https://doi.org/10.3389/fphar.2018.00146

Lipton, Z. C. (2016). The mythos of model interpretability. *ArXiv, 1606.03490*. Retrieved from http://arxiv.org/abs/1606.03490

Liu, J., Mansouri, K., Judson, R. S., Martin, M. T., Hong, H., Chen, M., ... Shah, I. (2015). Predicting hepatotoxicity using ToxCast in vitro bioactivity and chemical structure. *Chemical Research in Toxicology, 28*, 738–751. https://doi.org/10.1021/tx500501h

Low, Y., Uehara, T., Minowa, Y., Yamada, H., Ohno, Y., Urushidani, T., ... Tropsha, A. (2011). Predicting drug-induced hepatotoxicity using QSAR and toxicogenomics approaches. *Chemical Research in Toxicology, 24*(8), 1251–1262. https://doi.org/10.1021/tx200148a

Low, Y. S., Alves, V. M., Fourches, D., Sedykh, A., Andrade, C. H., Muratov, E. N., ... Tropsha, A. (2018). Chemistry-Wide Association Studies (CWAS): A novel framework for identifying and interpreting structure-activity relationships. *Journal of Chemical Information and Modeling, 58*(11), 2203–2213. https://doi.org/10.1021/acs.jcim.8b00450

Luo, C., Wu, D., & Wu, D. (2017). A deep learning approach for credit scoring using credit default swaps. *Engineering Applications of Artificial Intelligence, 65*, 465–470. https://doi.org/10.1016/j.engappai.2016.12.002

McCarthy, J., Minsky, M., Rochester, N., & Shannon, C. (1955). *A proposal for the Dartmouth summer research project on artificial intelligence*. Retrieved December 4, 2019, from http://www-formal.stanford.edu/jmc/history/dartmouth/dartmouth.html

Melo Calixto, N., Braz dos Santos, D., Clecildo Barreto Bezerra, J., & de Almeida SilvaID, L. (2018). In silico repositioning of approved drugs against Schistosoma mansoni energy metabolism targets. *PLoS One*. https://doi.org/10.1371/journal.pone.0203340

Melo-Filho, C. C., Dantas, R. F., Braga, R. C., Neves, B. J., Senger, M. R., Valente, W. C. G., ... Andrade, C. H. (2016). QSAR-driven discovery of novel chemical scaffolds active against Schistosoma mansoni. *Journal of Chemical Information and Modeling, 56*(7), 1357–1372. https://doi.org/10.1021/acs.jcim.6b00055

Miotto, R., Wang, F., Wang, S., Jiang, X., & Dudley, J. T. (2017). Deep learning for healthcare: Review, opportunities and challenges. *Briefings in Bioinformatics, 19*(6), 1236–1246. https://doi.org/10.1093/bib/bbx044

Mitchell, T. M. (1997). *Machine learning*. New York, NY: McGraw-Hill.

Neves, B. J., Braga, R. C., Alves, V. M., Lima, M. N. N., Cassiano, G. C., Muratov, E. N., Costa, F.T.M., Andrade, C. H. (2019). Deep Learning-driven research for drug discovery: Tackling Malaria. *PLOS Computational Biology, 16*(2):e1007025, https://doi.org/10.1371/journal.pcbi.1007025

Neves, B. J., Dantas, R. F., Senger, M. R., Melo-Filho, C. C., Valente, W. C. G., de Almeida, A. C. M., ... Andrade, C. H. (2016). Discovery of new anti-schistosomal hits by integration of QSAR-based virtual screening and high content screening. *Journal of Medicinal Chemistry, 59*(15), 7075–7088. https://doi.org/10.1021/acs.jmedchem.5b02038

Nosengo, N. (2016). Can you teach old drugs new tricks? *Nature, 534*(7607), 314–316. https://doi.org/10.1038/534314a

Pantaleao, S. Q., Fujii, D. G. V., Maltarollo, V. G., da C. Silva, D., Trossini, G. H. G., Weber, K. C., … Honorio, K. M. (2017). The role of QSAR and virtual screening studies in type 2 diabetes drug discovery. *Medicinal Chemistry, 13*(8), 706–720. https://doi.org/10.2174/1573406413666170522152102

Perols, J. (2011). Financial statement fraud detection: An analysis of statistical and machine learning algorithms. *Auditing: A Journal of Practice & Theory, 30*(2), 19–50. https://doi.org/10.2308/ajpt-50009

Ping, P., Watson, K., Han, J., & Bui, A. (2017). Individualized knowledge graph: A viable informatics path to precision medicine. *Circulation Research, 120*(7), 1078–1080. https://doi.org/10.1161/CIRCRESAHA.116.310024

Polishchuk, P., Kuz'min, V., Artemenko, A., & Muratov, E. (2013). Universal approach for structural interpretation of QSAR/QSPR models. *Molecular Informatics, 32*, 843–853.

Renard, P., Alcolea, A., & Ginsbourger, D. (2013). Stochastic versus deterministic approaches. In J. Wainwright & M. Mulligan (Eds.), *Environmental modelling: Finding simplicity in complexity* (2nd ed.). Chichester, UK/Hoboken, NJ: Wiley.

Ruths, D., & Pfeffer, J. (2014). Social media for large studies of behavior. *Science, 346*(6213), 1063–1064. https://doi.org/10.1126/science.346.6213.1063

Speck-Planche, A. (2019). Multicellular target QSAR model for simultaneous prediction and design of anti-pancreatic cancer agents. *ACS Omega, 4*(2), 3122–3132. https://doi.org/10.1021/acsomega.8b03693

Sushko, I., Novotarskyi, S., Körner, R., Pandey, A. K., Cherkasov, A., Li, J., … Tetko, I. V. (2010). Applicability domains for classification problems: Benchmarking of distance to models for Ames mutagenicity set. *Journal of Chemical Information and Modeling, 50*(12), 2094–2111. https://doi.org/10.1021/ci100253r

Tildesley, D., & Care, P. (2014). *Press release: Next RSC president predicts that in 15 years no chemist will do bench experiments without computer-modelling them first.* Retrieved from http://www.rsc.org/AboutUs/News/PressReleases/2013/Dominic-Tildesley-Royal-Society-of-Chemistry-President-Elect.asp

Todeschini, R., & Consonni, V. (2009). *Molecular descriptors for chemoinformatics* (R. Mannhold, H. Kubinyi, & G. Folkers, Eds.). https://doi.org/10.1002/9783527628766

Tropsha, A. (2010). Best practices for QSAR model development, validation, and exploitation. *Molecular Informatics, 29*(6–7), 476–488. https://doi.org/10.1002/minf.201000061

Vamathevan, J., Clark, D., Czodrowski, P., Dunham, I., Ferran, E., Lee, G., … Zhao, S. (2019). Applications of machine learning in drug discovery and development. *Nature Reviews Drug Discovery, 18*(6), 463–477. https://doi.org/10.1038/s41573-019-0024-5

Wang, Y., Xiao, J., Suzek, T. O., Zhang, J., Wang, J., Zhou, Z., … Bryant, S. H. (2012). PubChem's BioAssay database. *Nucleic Acids Research, 40*(Database issue), D400–D412. https://doi.org/10.1093/nar/gkr1132

Xu, C., Cheng, F., Chen, L., Du, Z., Li, W., Liu, G., … Tang, Y. (2012). In silico prediction of chemical Ames mutagenicity. *Journal of Chemical Information and Modeling, 52*(11), 2840–2847. https://doi.org/10.1021/ci300400a

Zhang, L., Fourches, D., Sedykh, A., Zhu, H., Golbraikh, A., Ekins, S., … Tropsha, A. (2013). Discovery of novel antimalarial compounds enabled by QSAR-based virtual screening. *Journal of Chemical Information and Modeling, 53*(2), 475–492. https://doi.org/10.1021/ci300421n

Zhang, S., Wei, L., Bastow, K., Zheng, W., Brossi, A., Lee, K. H., & Tropsha, A. (2007). Antitumor agents 252. Application of validated QSAR models to database mining: Discovery of novel tylophorine derivatives as potential anticancer agents. *Journal of Computer-Aided Molecular Design, 21*(1–3), 97–112. https://doi.org/10.1007/s10822-007-9102-6

Zhao, K., & So, H.-C. (2019). Using drug expression profiles and machine learning approach for drug repurposing. *Methods in Molecular Biology (Clifton, N.J.), 1903*, 219–237. https://doi.org/10.1007/978-1-4939-8955-3_13

Zhu, X., & Kruhlak, N. L. (2014). Construction and analysis of a human hepatotoxicity database suitable for QSAR modeling using post-market safety data. *Toxicology, 321*(1), 62–72. https://doi.org/10.1016/j.tox.2014.03.009

Chapter 8
Conclusion

Abstract A summary is presented of the concluding observations from previous chapters in which it is noted that complexity underpins chemistry, physicochemical properties, pharmaceutical manufacturing, preclinical observations, and population biology (of clinical relevance) and has implications for machine learning and artificial intelligence. For each of these, topic tools or mathematical interpretations have been presented. The concluding observation for the volume, as it appeared in the first edition, appears to be somewhat prescient. Rather than the typical data-to-wisdom overlay that has been placed on ontologies emerging from the so-called big data revolution, it is clear that there are feedback elements from wisdom to knowledge and information such that rather than drawing heuristic inferences from existing data, we can take a holistic view of the entire framework to allow the most meaningful interpretation. This seems to be a key characteristic of all data analytic approaches that are underway in science and society currently.

Keywords Complexity · Physics · Chemistry · Quality · Performance · Pharmacokinetics · Pharmacodynamics · Population biology · Data management

The complexity observed in pharmaceutical systems that guide medical diagnosis through therapeutic agent discovery to the patient is a microcosm of that observed in nature. We rarely reflect on the barriers that Linnaeus addressed in developing a taxonomy for the known biological world or the Mendeleev overcame in developing the periodic table of chemicals (Ghosh & Kiparsky, 2019; Paterlini, 2007).

© American Association of Pharmaceutical Scientists 2020
A. J. Hickey, H. D.C. Smyth, *Pharmaco-complexity*, AAPS Introductions in the Pharmaceutical Sciences, https://doi.org/10.1007/978-3-030-42783-2_8

Both were faced with enormous amounts of data that had not been organized into an understandable system of functionality.

As the twenty-first century opened, we were suddenly (of course after decades, if not centuries, of foundational research) faced with multiple large databases. The promise of exploring these resources was both exciting and intimidating. The availability of extraordinary computing capacity and the support of vast numbers of data and information scientists make this a unique historical moment. A revolutionary change in biomedical science is imminent, and we should embrace the challenge with the enthusiasm that guided previous generations.

It should be no surprise to pharmaceutical scientists and engineers that they work in an arena of complex multivariate phenomena which require costly and time-consuming efforts to monitor, understand, predict, and control. Indeed, it might be argued that an outcome of this small volume is to state the obvious. However, the approaches taken to maneuver in order to gain as much knowledge as possible if the systems involved to integrate them into larger product development and therapeutic objectives are dictated by the ability to acquire data reflecting the actual nature of the situation and interpret it appropriately.

In Chap. 1, the concepts of mathematical complexity consisting of tools to address data were introduced briefly. The more philosophical was raised of the way in which we communicate our findings to improve products and therapies and the hierarchical bias that we often use to depict thinking on this topic. Moreover, the complex ways in which fields "overlap" do not fully capture, from an intellectual perspective, the integrative nature of each of these elements. It was proposed that the reflections of the following chapters might lead us to a slightly different way of considering these issues.

Chapter 2 introduces the notion that complexity in the physics and chemistry underpinning important pharmaceutical principles is not a new observation. However, it should be noted that the apparent simplicity of our interpretation of molecular dynamics and chemical kinetics belies a level of complexity that we have yet to fully probe, and this may be of future importance. The evolution of understanding, over approximately a century, of aspects of molecular association and surface chemistry is described as a foundation for considering recent developments. It is evident as we scrutinize systems more closely that they become susceptible to a level of understanding that in the absence of current analytical and probing methods would not be possible. In short, increasing the amount of available data allows for more accurate interpretation of observations, and ultimately this knowledge will facilitate prediction and control.

Chapter 3 describes the importance of complex physicochemical properties and methods of interpreting them with respect to the performance of solid dosage forms which represent the majority of pharmaceutical products. The convergence of mathematical complexity and the true nature of these systems has been one of the intriguing occurrences in the last two decades, as it points to the potential for a level of prediction and control that has hitherto been unimaginable.

Chapter 4 describes the considerations that are uppermost in pharmaceutical manufacturing processes. This topic is of great concern to scientists, engineers, and regulators trying to bring the highest level of control to the quality and performance and thereby serving patients with reliable and therapeutic products. The mathematical approaches to this have focused predominantly on statistical process control. It might be proposed that the number of unknown variables, essential to the complete nature of the process, is not or cannot be known. Consequently, the most efficient approach, at this time, is to isolate the variables understood to contribute most significantly to the process and to mitigate against unknown environmental or time-dependent variables. However, there are fields in mathematical complexity that would allow an intelligent approach wherein the analytical system "learns" about the process and brings that knowledge to bear on control. Notably, the field of artificial neural networks has seen the greatest interest for this purpose and presumably will continue to evolve to suit the product development needs.

In order to do justice to the scope of product development, clinically important parameters of pharmacokinetics and pharmacodynamics were addressed in Chap. 5. It is clear that the nature of the biological response to the administration of a therapeutic (or preventative) agent is as complicated as the dosage form and its production. These topics have historically been approached with simplifying assumptions, but the capacity to develop physiologically and pharmacologically relevant systems approaches indicates the great potential to address Ehrlich's goal of the "magic bullet" (Albert, 1960).

The impact of a full understanding of population biology on approaches to product development is necessarily speculative as described in Chap. 6. Since the last decade can, without hyperbole, be considered the dawn of a new age in biology, much has yet to be done. Nevertheless, our current knowledge of genomics, proteomics, and metabolomics indicates that we will need complex analyses as presaged by the rising interest in pharmacoinformatics. It is likely that research in biology will allow specific drug targeting within populations and subpopulations and possibly ultimately at the level of routine individualized medicine. This will, undoubtedly, feedback to the earlier stages of product development requiring speed and flexibility and much greater understanding of the processes involved in order to continue to ensure product quality and performance.

Reflecting on the conception of the way in which we manage data to gain understanding presented in Chap. 1, and in light of the multiple complex processes involved in product development, a more general scheme can be proposed. If we assume that there are existing domains in the realms of understanding, i.e., data, information, knowledge, and wisdom as shown in Fig. 8.1, arguably these might be considered infinite, in which we are not aware of the boundaries. Then we can map one domain onto another, and there are reflections which occur from "higher" to "lower" levels. Exploring the relationships within not only one process but interfaces or integrations into others may make progress in treatment and prevention of disease more rapid and advance our understanding of "the big picture." Undoubtedly, from a practical standpoint, this can only be achieved by the application of computers, their increased processing capacity, and information technology. Whether, as has been suggested elsewhere, we need to invoke sentient computers to fully explore the domains identified in Fig. 8.1 is beyond the scope of this text but is an interesting philosophical question (Tipler, 1994).

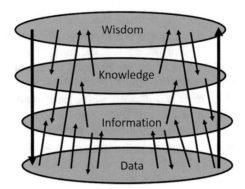

Fig. 8.1 Schematic depiction of the way in which data impinge on understanding (information, knowledge, and wisdom) indicating the proposition that the mapping occurs through domains which represent "all" of the items indicated and consequently imply a link between all relevant phenomena that ultimately, in an ideal world, allows prediction and control of any phenomenon

Fig. 8.2 Schematic depiction of the complex relationships between the selected processes described (PE, pharmaceutical engineering (manufacturing); DF, dosage form; PC, physicochemical factors; IP, individual pharmacokinetics and pharmacodynamics; and PB, population biology)

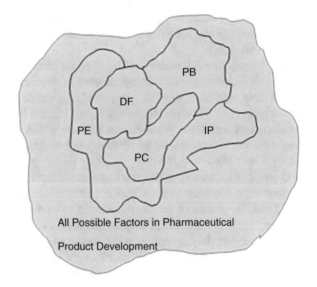

Taking the general principles of knowledge management and returning to the conception of the relationship between the processes discussed in this volume, it is clear that there is a domain of all possible factors involved in product development, and in order to predict and control the ones we can clearly define (and hopefully others that might be identified in the future), their complex interactions must be explored to understand the continuum from drug molecule to disease treatment or prevention (Fig. 8.2). Meta-analysis is now feeding back to data generation in every field. As data management has evolved and repositories and registries have become common, it is clear that there is a requirement for greater clarity with respect to methods and controls for data generation if specific applications are to be adequately illuminated. Chapter 7 was added to this edition to reflect the enormous advances that have occurred in data management, tool development, and machine learning that have supported pattern recognition promoting rapid diagnosis and therapy.

The goal of this outline has been to present the practical implications of pharmaco-complexity in selected scientific areas of importance to pharmaceutical product development and in turn to the pharmaceutical scientist. In presenting the factual background, an attempt has been made to raise more philosophical questions regarding the way we approach these complex phenomena in order to gain the level of understanding required to guarantee the quality and performance of products. If this encourages greater interdisciplinary thinking on the nature of future therapy, this will justify the effort expended.

The first edition of this short primer appeared as the scientific community was beginning to engage with the breadth of data and complex interactions that were emerging from the convergence of molecular biology, computer technology, and the Internet. A decade has passed, and we are now immersed in curated data; electronic methods of data collection and analysis; the promise of machine learning and artificial intelligence; the ability to do rapid whole human genome sequencing (Guinness World Record held by Rady Children's Institute, San Diego, CA, 19 hours); high-throughput transcriptomics, proteomics, and metabolomics; and the capacity to discover and develop new therapeutics in record times. However, despite the new science and methodology based on acknowledgment of the pharmaco-complexity, some areas are lagging. Commercial development and regulation continue to be based on twentieth-century rationales and are constrained by archaic approaches that generally result in unnecessarily slow development and approval times and unacceptably expensive products. This is not a criticism of those engaged in this endeavor; rather it reflects the disconnect between the data and communication revolution and the ability of the physical world of products and services to adapt. The diligent efforts of the pharmaceutical industry and regulatory agencies to accelerate their activities and remove barriers are commendable. We can only hope that events will precipitate disruptive change and that all patients can expect to have their disease ameliorated or cured through comprehensive engagement with the science and technology of pharmaco-complexity.

References

Albert, A. (1960). *Selective toxicity: The physico-chemical basis of therapy* (2nd ed.). New York, NY: Wiley.

Ghosh, A., & Kiparsky, P. (2019). The grammar of the elements. *American Scientist, 107*(6), 350.

Paterlini, M. (2007). There shall be order. The legacy of Linnaeus in the age of molecular biology. *EMBO Reports, 8*(9), 814–816.

Tipler, F. (1994). *The physics of immortality*. New York, NY: Anchor Books Doubleday.

Index

A
Artificial intelligence (AI), 3, 59, 60, 62, 74

B
Batch production methods, 42
Brunauer Emmett Teller (BET)
      theory, 12

C
Carr's compressibility index, 28
Catastrophe, 14
Central composite design (CCD), 40
Classical Emax model, 49
Critical quality attributes (CQA),
     37, 42

D
Data, information, knowledge, and
     wisdom, 73
Databases, 48, 62

H
Hartley model, 9
Hixon-Crowell cube root law, 23
Homeostatic mechanisms, 54

K
Knowledge management, 55, 74

L
Langmuir's adsorption equation, 11
Latin square, 39

M
Machine learning, 3, 4, 59–64, 74
Mass action law, 10, 11, 49
Micelles, 8–11, 63
Mycobacterium tuberculosis, 55

N
New chemical entities (NCEs), 21, 50
Noyes-Whitney equation, 23

P
Pharmaceutical manufacturing, 40
    definition, 37, 43
    process design and control, 40–42
        batch production methods, 42
        design space development, 41, 42
        quality, 40–42
        quantitative analytical methods, 42
    risk assessment and management, 42–43
    statistics and experimental design, 38–40
        CCD, 40
        factorial design, 39, 40
        fractional factorial design, 39
        Latin square designs, 39
        response surface maps, 40, 41
        sources of error, 38

© American Association of Pharmaceutical Scientists 2020
A. J. Hickey, H. D.C. Smyth, *Pharmaco-complexity*, AAPS Introductions in the
Pharmaceutical Sciences, https://doi.org/10.1007/978-3-030-42783-2

Printed in the United States
by Baker & Taylor Publisher Services